SCIENTIFIC UNCERTAINTY
AND THE POLITICS OF WHALING

SCIENTIFIC UNCERTAINTY
AND THE POLITICS OF WHALING

Michael Heazle

UNIVERSITY OF WASHINGTON PRESS Seattle and London
CANADIAN CIRCUMPOLAR INSTITUTE (CCI) PRESS Edmonton

THIS BOOK IS PUBLISHED IN MEMORY OF MARSHA L. LANDOLT
(1948–2004), DEAN OF THE GRADUATE SCHOOL AND VICE PROVOST,
UNIVERSITY OF WASHINGTON, WITH THE SUPPORT OF THE UNIVERSITY
OF WASHINGTON PRESS ENDOWMENT.

Copyright © 2006 by the University of Washington Press
Printed in the United States of America
12 11 10 09 08 07 06   5 4 3 2 1

All rights reserved. No part of this publication may be reproduced or transmitted in any form or by any means, electronic or mechanical, including photocopy, recording, or any information storage or retrieval system, without permission in writing from the publisher.

Library of Congress Cataloging-in-Publication Data
Heazle, Michael.
Scientific uncertainty and the politics of whaling /
Michael Heazle.—1st ed.
p. cm.
Includes bibliographical references and index.
ISBN 0-295-98605-0 (hardback : alk. paper)
1. Whaling—Political aspects. 2. International Whaling Commission.
3. Whales—Conservation—Decision making. 4. Uncertainty
(Information theory)   I. Title.
SH381.H42 2006   333.95'95—dc22   2006004334

Library and Archives Canada Cataloguing in Publication
Heazle, Michael
Scientific uncertainty and the politics of whaling / Michael Heazle.
(Circumpolar research series, ISSN 0838-133X ; no. 11)
Co-published by the Canadian Circumpolar Institute (CCI) Press.
Includes bibliographical references and index.
ISBN 0-295-98605-0 (UWP)
ISBN 1-896445-37-3 (CCIP) Circumpolar Research Series No. 11;
Studies in Whaling No. 8
1. International Whaling Commission. 2. Whaling—Political aspects.
3. Whales—Conservation—Decision making. 4. Uncertainty (Information theory).   I. Title. II. Series.
SH381.H42 2006a   333.95'95   C2006-901321-7

The paper used in this publication is acid-free and 90 percent recycled from at least 50 percent post-consumer waste. It meets the minimum requirements of American National Standard for Information Sciences—Permanence of Paper for Printed Library Materials, ANSI Z39.48–1984.♾

For Kazu, Alexi, and Jake

CONTENTS

       Acknowledgments   ix
       List of Abbreviations   xi

ONE   Introduction   3

TWO   The IWC 1949–59: An Exercise in Uncertainty Becoming Certainty   36

THREE   The Antarctic Collapse: Uncertainty Takes a (Brief) Holiday   66

FOUR   The Worm Turns: The IWC's Reinterpretation of Uncertainty   107

FIVE   Scientific Uncertainty and the Evolution of the Superwhale   133

SIX   Conclusion   177

       Appendix   189
       Notes   203
       Bibliography   237
       Index   253

ACKNOWLEDGMENTS

In the course of completing this work, I have been fortunate to have had the help and cooperation of a great many people—all of whom kindly gave their time and effort to answer my many questions. Firstly, I would like to thank Dr. John Butcher and Professor Nick Knight, my colleagues at Griffith University, for all their generous advice, support, and patience. Both always were ready and willing to give excellent guidance and advice, moral support, and a great deal of their time to help out in whatever way they could. I also would like to acknowledge the invaluable help I received from Dr. William Aron, Dr. Ray Gambell, and Professor Milton Freeman. Each very kindly read various drafts and gave excellent comments and advice that, in addition to always being insightful and clearly explained, made my efforts at charting a course through the many events and issues that make up the IWC's history far more manageable.

I also am indebted to the numerous other people who spent some of their valuable time answering my e-mails and, in many cases, meeting with me personally to discuss my research and answer questions. In this regard, there are many to whom I am greatly obliged, but I would like to mention in particular Greg Donovan, Martin Harvey, Professor Doug Butterworth, John Bannister, Dr. Peter Best, Dr. Jeff Breiwick, and Dr. Doug DeMaster in order to express my deep appreciation for their very kind interest in my research.

A special vote of thanks also is due to the staff at the IWC office in Histon and the Institute of Cetacean Research in Tokyo who kindly made their libraries available to me and helped track down many important sources. Julie Creek and Cherry Allison at the IWC office, and also Mitsuyoshi Murakami and Dan Goodman at the Institute of Cetacean Research, were all more than helpful in this regard and to each I offer a very sincere "thank you." The IWC and University of California Press both have allowed me to adapt and reproduce tables and maps from their publications in this work, and I would like to acknowledge their generous support. Equally appreciated is the excellent support I received from the Griffith Asia Institute in preparing the final manuscript, in particular from Professor Michael Wesley and Robyn White, and also from the editorial and production staff at the University of Washington Press; I especially would like to thank Jacqueline Ettinger for all her wonderful support and guidance.

There are numerous other people who also deserve mention, but unfortunately space does not allow for this here. However, the names of all the people who kindly took time to answer my questions, either through correspondence or interviews, are listed in the bibliography. Needless to say, any omissions or errors that remain in this study are my own.

Last, but certainly not least, I would like to say a very big "thank you" (and "sorry") to my wife Kazu and our two children, Alexi and Jake, for putting up with me and for being so supportive of my efforts to research and write this book. Without their great patience, love, and support, none of this would have been possible.

MICHAEL HEAZLE
December 2005

LIST OF ABBREVIATIONS

AHSC	Ad Hoc Scientific Committee
BWU	Blue Whale Unit
CLA	catch limit algorithm
COF	Committee of Four
COT	Committee of Three
CPUE	catch per unit effort
CV	coefficient of variation
FAO	UN Food and Agriculture Organization
ICRW	International Convention for the Regulation of Whaling
IOS	International Observer Scheme
IWC	International Whaling Commission (called the International Commission on Whaling until 1975)
MSYL	maximum sustainable yield level
MSY	maximum sustainable yield
NGO	nongovernmental organization
NMP	New Management Procedure
RMP	Revised Management Procedure
RMS	Revised Management Scheme
SC	Scientific Committee
TC	Technical Committee

SCIENTIFIC UNCERTAINTY
AND THE POLITICS OF WHALING

1
INTRODUCTION

Established under the 1946 International Convention for the Regulation of Whaling (ICRW), the International Whaling Commission (IWC) has, since its first meeting in 1949, been the main forum for debate over the hunting of various whale species and the kinds of management that should be applied to them. Since its inaugural meeting, however, the IWC has struggled to create and implement a regime capable of managing both whales and whalers in a sustainable manner. Fifty years on, the commission's attempts at compromise between the needs of the whaling industries and its responsibility to protect whale stocks from depletion continue to be undermined by a dearth of agreement among IWC members over the commission's goals and how they should be pursued—thereby creating a situation that bodes ill not only for the future of both the IWC and the animals it is intended to manage, but also for the future of international cooperation on wildlife management problems and issues at large.

At the heart of the IWC's ongoing management impasse has been the double-edged role played by scientific uncertainty in the commission's policy-making process. This study focuses on this issue to explore why the commission has been divided and frustrated in its attempts to agree on how and to what ends whales should be managed. Thus, in terms explaining why the commission's pro- and anti-whaling blocks remain locked in stalemate over the moratorium, the single most important development within the IWC over the last fifty years was a dramatic switch in how many formerly pro-whaling members perceived scientific uncertainty—a gestalt-like change in perception concerning how scientific uncertainty should be understood—which occurred incrementally between the collapse of the Antarctic stocks in the early 1960s and the moratorium's eventual adoption in 1982.

In the 1950s and early 1960s, scientific advice urging lower catch limits was opposed by the then majority of active whaling governments in the commission on the grounds that there was too little scientific evidence, and therefore too much uncertainty, to support contentions that blue, fin, and humpback whale stocks were being depleted. The early 1960s, however, brought the realization that uncertainty had been overplayed in the rejection of lower quotas, as the Antarctic stocks collapsed along with the commercial viability of pelagic whaling for all but two of the five nations involved. The result was that during the latter half of the 1960s only Soviet and Japanese whaling in the Antarctic continued. Soviet and Japanese catches, however, were restricted by the IWC's now much smaller quotas and complete protection of two of the whaling industry's three traditionally preferred species (the blue and humpback) by the mid-1960s. The third whale of choice, the fin, was hunted until 1975–76, but only at greatly reduced levels compared with the 1950s and early 1960s due primarily to the increasing scarcity of this species. In response, the Japanese and Soviet fleets concentrated instead on the smaller sei and minke whales in order to continue whaling.

Then, by the early 1970s, essentially the same problems concerning scientific uncertainty over population numbers and sustainable catch limits again surfaced in the IWC. But this time, scientific uncertainty was used, primarily by the United States, to argue in favor of a moratorium on all commercial hunting.[1] With strong support from the United States and a steadily growing number of other former whaling members, in addition to a contingent of new IWC members that had little or no experience with whaling, the moratorium was finally adopted by the IWC in 1982 and is yet to be lifted. Ironically, a number of the governments now citing scientific uncertainty in support of the moratorium previously had opposed reducing catches during the 1950s on the same grounds.

This curious switch within the IWC concerning the status of whales and how scientific uncertainty is interpreted has the potential, I believe, to explain a great deal about the commission's ongoing stalemate over its management goals in general and concerning the ends to which scientific advice has been used by the various groups and governments involved in the whaling debate. Thus, the central problem this study will focus on is the issue of why this switch in attitudes within the IWC occurred and what it can tell us about the current impasse over the moratorium.

Indeed, so far the IWC experience with attempting to regulate whaling has shown that the use of science as the basis for policy making—as called for in the IWC's charter, the ICRW—is by no means a simple or straightforward means of establishing a consensus-based and, therefore, workable environmental management regime, especially when seeking to manage a high-profile issue such as whaling. All of the IWC's major management debates to date—the struggle for lower catch limits in the 1950s and 1960s, the creation of the New Management Procedure (NMP) in 1974, the adoption of the moratorium in 1982, the commission's eventual acceptance in 1994 of the Revised Management Procedure (RMP) and the subsequent delays to its implementation—have involved arguments citing different perspectives on scientific uncertainty.

The issue of scientific uncertainty and its treatment within the commission, therefore, provides a thread of continuity with which the commission's policy changes and accompanying interaction between political objectives and scientific advice can be analyzed and, I hope, better understood. By attempting to explain why and how scientific uncertainty was first used to oppose reduced catches and then was later reinvoked by many of the same members to support zero catches (i.e., the moratorium), some new light will be shed not only on why the IWC has become trapped in its current impasse over the moratorium, but also on the reasons why perceptions of scientific uncertainty change and the impact such changes can have on wildlife management.

This study's examination of scientific uncertainty in the IWC is limited to the commission's management of baleen whales in the Antarctic for several reasons. Firstly, a detailed investigation of the IWC's management of all species and stocks would require more voluminous and extensive research than is appropriate for the purposes and objectives of this study. Secondly, baleen whaling in the Antarctic, for most of the twentieth century and the first two decades of the IWC's existence, was by far the largest source of catches for the whaling industry. Consequently, the majority of the IWC's management decisions and debates prior to the moratorium's adoption in 1982 focused upon the exploitation of the Antarctic's once numerous baleen whales.

Whaling in the North Pacific, the next most important region for coastal and pelagic whaling, consisted mostly of only small-scale Soviet and Japanese catches during the 1950s and did not increase significantly against Antarctic catches until the late 1960s, following the collapse of the Antarctic stocks. Until 1970, when the IWC adopted quotas for North Pacific stocks for the first time (North Pacific minke whale quotas were not established until 1972), the Antarctic remained the only region where whaling was regulated by annual IWC quotas. Another six years would pass before the IWC finally established quotas for all exploited stocks worldwide.[2] Thirdly, the exploitation of sperm whales—a toothed rather than baleen whale[3]—is not dealt with in this study primarily because these catches were small in comparison to the number of baleen whales taken, but also because the inedible nature of sperm-whale oil (and the general unpopularity of sperm-whale meat) and its use primarily as a high-grade lubricant meant that sperm-whale catches were part of an industry that was very different from that which exploited the baleen stocks.[4]

Questioning the role and influence of scientific advice in policy making is by no means a new academic pursuit. But many questions concerning how environmental policy makers interpret scientific research and advice remain unanswered, particularly in relation to the IWC, where much of the commentary largely has ignored the question of why policy makers often interpret scientific uncertainty differently. Although a great deal has been written about the IWC and its policies—both in the media and in academic literature—much of this material has been highly polemic and has either accepted the role of science as a basis for policy making in an unquestioning or superficial manner or has ignored issues and problems related to science altogether.[5] The animal rights and environmental ethics literature related to whaling falls into the latter category, as does the voluminous legal analysis of the IWC and its policies; much of the international relations literature largely falls into the former. Animal rights and legal discussions of the IWC are, for the most part, beyond the focus of this work and therefore will not be discussed in any detail except on those occasions where there exists some relevance to my analysis of the IWC's treatment of scientific uncertainty.

Not surprisingly, most of the existing work on the evolution and function of international environmental regimes—such as the IWC's International Convention on the Regulation of Whaling, the Convention on International Trade in Endangered Species of Wild Flora and Fauna, the International Agreement on the Conservation of Polar Bears, and the Montreal Protocol on Ozone Layer Depletion—has been provided by a broad scope of international relations theorists and researchers focusing largely on regime theory. Within the broadly defined field of regime theory, which includes game theory, functionalism, structuralism, and cognition among its main theoretical approaches, various arguments have been constructed over the last two decades to counter the traditional realist view of international relations, which largely has dominated international relations theory.[6] And while the theoretical efforts of most international relations researchers are also outside the scope of this study, the work of cognitivist theorists, most notably Peter Haas,[7] and their development of the concept of epistemic communities will be discussed in the conclusion—in the light of my own historical analysis of scientific uncertainty in the IWC—because of the central role this approach gives to scientific advice in its explanation of why regimes matter and how they can work.

In the case of the IWC, and the broader field of environmental policy, significant problems concerning the treatment of scientific uncertainty in policy making have been raised by the relatively recent emergence of the so-called precautionary principle. The precautionary principle has generated controversy and has attracted considerable attention from academics and policy makers, many of whom disagree over how the precautionary principle might be applied to particular environmental problems. But while critiques and commentaries concerning the precautionary principle as applied to environmental policy making have emerged over recent years, there remains little detailed discussion of its impact upon the IWC and the significant role it has played in the commission's drift toward redundancy—and of how the use of scientific uncertainty by IWC members might be better understood.

THE PRECAUTIONARY PRINCIPLE AND PERCEPTIONS OF SCIENCE

The precautionary principle is essentially a loose set of guidelines intended to help policy makers manage scientific uncertainty in the course of developing environmental policies and to avoid consequences science may be unable to foresee. The principle's fundamental purpose is to encourage the adoption of policies that reduce the risk of environmental damage by erring on the side of caution, and its use has significant implications for existing scientific method and the kind of conclusions research produces on environmental issues—particularly in the IWC, where the precautionary principle frequently has been invoked to argue against scientific advice that supports any return to commercial whaling.

The precautionary principle's emphasis on what may happen as a result of a given activity in effect magnifies the importance of uncertainty by demanding evidence that something will not happen as opposed to the conventional scientific approach of describing the likelihood of a future outcome in terms of existing evidence. Put another way, scientific methods traditionally have demanded that conclusions about possible outcomes for which little or no reliable empirical evidence exists should be discarded as speculation. The precautionary principle, however, challenges this approach by rejecting the absence of evidence as sufficient reason for believing something is unlikely to happen and demanding the existence of evidence to support such conclusions.

Scientists normally assign a very low order of probability to events that no clear evidence indicates will happen, and their judgment in this regard is determined by the induction/deduction-based epistemology upon which conventional scientific reasoning is based. Thus, the precautionary principle questions the veracity of science on its most fundamental level and, by doing so, also brings into question science's reliability in identifying and predicting possible outcomes. A recent experiment by U.S. scientists, attempting to re-create the birth of the universe by slamming subatomic particles into each other at 99.995 percent of the speed of light, provides an excellent example of the standard scientific response to potential problems raised by a precautionary approach. When quizzed on the possibility that the experiment's goal of "generating temperatures of a trillion degrees and creating a substance that has not existed for thirteen billion years" might cause "some kind of catastrophe," a panel of physicists investigating the experiment's potential dangers reported, "Our conclusion is that the candidate mechanisms for catastrophe scenarios . . . are firmly excluded by existing empirical evidence, compelling theoretical arguments, or both."[8]

So despite fears that the American experiment could have destroyed life as we know it, and possibly even the entire universe, the re-creation went ahead (without any significant opposition nor any apocalyptic effects, as far as we know) entirely on the basis of scientific opinion informed only by inductive and deductive reasoning based on "existing empirical evidence." In this instance, scientific method and reasoning were deemed sufficient to discount the precautionary concerns of the media and, no doubt, some other scientists. But in other cases posing far less cataclysmic dangers, commercial whaling for example, similarly informed (i.e., inductive/deductive-based) scientific opinion has been vehemently argued against and has also been the focus of fierce public opposition. Thus, on the surface at least, it seems that the extent to which the precautionary principle is applied to a given situation may be determined by factors other than issues of science and knowledge alone.

The role of science in influencing the choices we make and the policies governments pursue is far from clear, in spite of the popular respect science still receives as the main source of human knowledge. A basis in science is still widely employed as the main criterion for separating truth and knowledge from myths, rumors, and old wives' tales. Practically everyone, from advertising executives and politicians to academics, seems to embrace the mantle of science in order to acquire the authority of "knowing," which is required in most modern societies if one is to "sell" his or her ideas to others. According to Alan Chalmers,

> the high regard for science is not restricted to everyday life and the popular media. It is evident in the scholarly and academic world too. Many areas of study are now described as sciences by their supporters, presumably in an effort to imply that the methods used are as *firmly based* and as potentially *fruitful* as in a traditional science such as physics or biology. Political science and social science are by now common place. Many Marxists are keen to insist that historical materialism is a science.[9]

But the evident high regard of modern societies for the conventional notion of science and scientific practice sits uncomfortably with the growing popularity of the precautionary principle and the direction of environmental policies in general. The IWC, for example, on the one hand recognizes the veracity of science in its charter, the ICRW, which states that the commission's policies "shall be based on scientific findings,"[10] and its Scientific Committee (SC) spends a great deal of time gathering scientific research seemingly for this purpose. On the other hand, many of the IWC's major policy decisions over the last fifty years have been contrary to the majority advice of its SC and based largely on arguments against the reliability of scientific research. Why has this conflict over the role of science in policy making occurred? How can we account for the lack of controversy over scientific findings and opinions in some cases, but the preponderance of it in others, in particular, cases concerning environmental issues? Why is the precautionary principle approach to dealing with scientific uncertainty applied most vociferously in environmental policy debates, and what does this practice tell us about the IWC and its policies? And, conversely, what does the IWC's experience with attempting to manage whales on the basis of scientific advice tell us about the nature of science and its uses as a tool for policy making?

As we will see, the two major topics of this study—the IWC and scientific uncertainty—share an almost symbiotic relationship in so far as each is a useful and illustrative vehicle with which to study the other. By analyzing the IWC's major policy decisions over the last five decades and the commission's treatment of science in arriving at those policies, I will, in the first instance, construct a history of scientific uncertainty in the IWC. This account focuses on the earlier mentioned gestalt-like switch in the perception of whales that occurred within the IWC's membership, highlighting the role scientific uncertainty has played in preventing broad international cooperation on the management of whales.

The other side of my analysis concerns what the IWC can tell us about science and its inevitable uncertainties. The IWC experience with developing environmental policies, I believe, is a useful and productive means of exploring the relationship between science and the precautionary principle and also the issues and problems involved with basing environmental policy decisions (beyond those of the IWC) on these two factors.

In order to define science and what it represents, we first turn to an overview of the major problems the philosophy of science literature has raised concerning attempts to locate a scientific method. Then follows my own argument, in outline, of why perceptions of scientific uncertainty have changed among the IWC membership and the impact this change has had on the IWC and its policies.

THE SEARCH FOR THE SCIENTIFIC METHOD: ANOTHER HOLY GRAIL?

Two fundamental tenets of my argument are the assumptions that (a) science and politics should not be discussed as separate entities; and (b) uncertainty is unavoidable due to the inability of science to establish universal truths.

There has long been a tendency, among both the public and also many academics and scientists, to view science as an entirely objective practice in the pursuit of knowledge, somehow isolated from the subjective political imperatives of prioritization and expediency. A commonly cited reason for why science is, or at least should be, independent of politics and its associated pressures is that science is about revealing "the truth" about reality—a reality that is assumed to exist independently of us, regardless of what we might think or say about it. Science, therefore, is popularly regarded as the benchmark against which all assertions about the world can be measured and determined as either true or false. With this reasoning, it is possible to justify or reject political positions and actions on the basis of what is true and certain. In other words, science and the knowledge it produces should be the ultimate guide to political behavior, but politics, of course, must never interfere with science lest we risk distorting the very truth upon which we depend for guidance in our affairs.

For this view of science, it is essential that we be able to distinguish real science from pseudoscience and, therefore, genuine knowledge from myth, since we otherwise would be unable to tell if political decisions were guided by factual knowledge of the real world or merely by false representations manufactured for purely political ends. However, this view of what science represents and how it should be understood—science as unadulterated by politics and values—has become problematic over the last seventy years, to the point where, at the very least, such an assumption should be received with the utmost reservation.

When considering the proposition that science and politics should be considered separately, two major problems become apparent. Firstly, the idea that science is value-neutral is particularly difficult to defend, especially if one considers the extent to which ideology and political preferences (i.e., values) influence how scientists are funded, the goals of their research, the problems they pursue, and the criteria by which their research is judged—not to mention the even more fundamental question of how scientists could observe and think about an event in a culturally neutral and unbiased way. The second problem is a consequence of the first. In what sense can we consider "scientific inquiry" as capable of explaining reality, given the problematic nature of asserting a pure or value-neutral science? And, equally related, how can such a "scientific method" be distinguished from other "unscientific" methods?

Concerning the nature of science and questions about what science represents, two basic issues seem to have been dominant: (a) the search for demarcation between "real science" and "pseudoscience," which occupied much of Karl Popper's work and led him to his falsification theory; and (b) the debate over whether the evolution of science has been a rational or logical process. This debate was initially sparked in the early 1960s by Thomas Kuhn's criticisms of Popper's falsificationist position in his landmark work, *The Structure of Scientific Revolutions*, and was further expanded by Imre Lakatos, a former student of Popper's, and also by the more radical views of Paul Feyerabend.[11] Lakatos shared Kuhn's criticism of Popper concerning the mismatch between the falsificationist methodology and the historical evidence of how science had developed.

This problem led both Lakatos and Kuhn to agree (from very different positions) that any identification of a methodology of science, capable of serving as a definition for demarcation purposes, should be consistent with the history of science.

A basic question at the heart of this debate is why people like Popper and Lakatos (among many others) have considered it so important for science to be rational. The most obvious answer is that, if science cannot be demonstrated to be a rational process (i.e., a logical method that is taking humanity closer to "the truth" about reality), there can be no reason for setting science apart from so-called myth and superstition—which according to skeptics such as Feyerabend is indeed the case. Thus, without the status of rationality, science is no more capable of generating genuine knowledge than voodoo or witchcraft.

But an equally important problem is that without rationality, in the sense outlined above, science cannot be regarded as an objective and value-neutral process capable of rising above political manipulation and of protecting freedom and truth from state control. So for both Lakatos and Popper, rejecting Feyerabend's contention that, according to Larvor, "scientific rationality is not an ideologically neutral magistrate for the marketplace of ideas" but only one of many "competing ideologies" is essential.[12] The need to clearly identify "knowledge" was a consequence of the political and ethical dangers Popper and Lakatos believed would otherwise dominate a world without recourse to rational justification of what can or cannot be regarded as "true."

The problems facing any view of science as a distinctly rational and value-neutral quest for knowledge, however, are manifold. In the first instance, the issue of how a direct correspondence between the realm of knowledge (i.e., thoughts, ideas, and propositions) and the realm of objects (i.e., reality and its associated phenomena) can be established remains unresolved in any definitive sense. Empiricism and rationalism, the two basic approaches to this problem, both presuppose that an independent reality (or realm of objects) exists and that it can be accessed either through experience (in the case of empiricism) or through reasoning (as postulated by rationalism).

However, in addition to the problem of demonstrating the existence of an independent reality awaiting description, the logic of these two epistemologies is further complicated by the question of how they can verify their claims independently of the basic criterion of truth on which they rest: experience in the case of empiricism, and reason in the case of rationalism. In this way, both the empiricist and rationalist positions suffer from essentially the same problem of relying upon circular arguments to justify their claims, since both concepts present themselves as the standard against which knowledge claims should ultimately be tested.[13]

The idea of a direct correspondence between the realm of objects and the realm of thought is made more difficult yet by our inability to be objective about any observation of what might be assumed to be an independent reality, due to the plethora of differing experiences, values, and assumptions that individuals accumulate and carry with them and use as reference points through their lives. The problems raised by the theory- or value-dependent way in which we observe the world fundamentally affect any concept of knowledge based on the conventional notion of science as founded on empirical observation and rational logic.

Indeed, if one accepts the reasonable assumption that all observation is (as Popper argued) necessarily theory-dependent, it becomes difficult to argue that either historians or scientists can ever approach their discipline "objectively." The conventional notion of science is that of a search for "truth." But as E. H. Carr has pointed out, "truth" is "a word which straddles both the world of fact and the world of value, and is made up of both."[14]

In addition to the formidable problems posed by the theory-dependent nature of observation and facts—and the not insignificant implications they hold for claims that science and politics should be seen as separate entities—the image of science as an entirely logical enterprise has also been seriously undermined by David Hume's questioning of one of science's most fundamental methodologies, namely its reliance upon induction in order to derive theories from facts. Hume's position was that, due to the basic problems facing empiricism and rationalism, inductive reasoning (i.e., using instances of observed facts to arrive at general laws and theories) has no rational basis: it is neither logical in the deductive or rationalist sense (induction is not logically defensible since we are unable to observe all possible instances), nor is it justifiable in the empiricist sense without resorting to a circular inductive argument.[15] In his essay "Is Science Superstitious?" Bertrand Russell summed up the epistemological foundations and shortcomings of modern science in the following way:

> A particular blend of general and particular interests is involved in the pursuit of science; the particular is studied in the hope that it may throw light upon the general. In the middle ages it was thought that, theoretically, the

particular could be deduced from general principles; in the Renaissance these general principles fell into disrepute, and the passion for historical antiquity produced a strong interest in particular occurrences. This interest, operating upon minds trained by the Greek, Roman and scholastic traditions, produced at last the mental atmosphere which made Kepler and Galileo possible. But naturally something of this atmosphere surrounds their work, and has travelled with it down to their present day successors. "Science has never shaken off its origin in the historical revolt of the later Renaissance. It has remained predominately an anti-rationalist movement, based upon a naïve faith. What reasoning it has wanted has been borrowed from mathematics, which is a surviving relic of Greek rationalism, following the deductive method. Science repudiates philosophy. In other words, it has never cared to justify its faith or to explain its meaning, and has remained blandly indifferent to its refutation by Hume."[16]

Past attempts at justifying science as genuine knowledge (as opposed to ideology or mysticism) have largely focused on unsuccessful applications of logic and empiricist thinking, most notably by logical positivists during the early 1900s, in order to get around the problems posed by Hume. None of these approaches was able to resolve Hume's problem and were further discredited, as Popper pointed out, by their failure to account for the theory-dependence of observation. Popper, however, turned the problem on its head by introducing falsification, rather than verification, as a means of demonstrating the rational nature of scientific research and, thereby, a means of separating it from what he regarded as nonscientific pursuits.

Popper first introduced falsificationism in *The Logic of Scientific Discovery*, first published in German in 1934. The idea's ensuing, albeit short-lived, success as a solution to the problem of demarcation demolished the semantic-based criteria proposed by logical positivists for judging statements as true, false, or meaningless. Popper's theory was revolutionary; it liberated scientists from achieving scientific status through proof, instead inviting scientists to develop theories and then disprove them. According to Popper, having one's theory disproved was not failure, but rather an example of scientific progress and achievement.

Popper's falsificationism holds that the real mark of a scientific theory is whether or not it makes clear and concise statements or predictions about the world that can be tested by observation. In other words, the main tenet of Popper's idea was that because theories can never be proven—due to Hume's seemingly irreconcilable problem of induction—all that remains is for theories to be adopted until they are ultimately disproved. Popper's acceptance of Hume's problem as a given—in contrast to Kant, the logical positivists, and others—was falsificationism's major strength, since it dispensed with the need to prove anything and instead extolled the virtues of criticism and conjecture.

In Popper's view, theories designed to be immune from criticism—as he charged was the case with Marxism and psychology—were nothing more than unscientific dogma. Their continual restructuring by proponents to account for possible instances of falsification meant that such theories' testable empirical content was replaced with untestable generalizations. According to Popper,

> I can therefore gladly admit that falsificationists like myself much prefer an attempt to solve an interesting problem by a bold conjecture, *even (and especially) if it soon turns out to be false*, to any recital of a sequence of irrelevant truisms. We prefer this because we believe that this is the way in which we can learn from our mistakes; and that in finding that our conjecture was false we shall have learnt much about the truth, and shall have got nearer the truth.[17]

Thus, through falsification, Popper was able to logically rationalize induction by moving from a singular observation statement in order to *disprove* a universal law or theory, rather than attempting the logically indefensible by trying to move from single statements in order to *justify* a law or theory. Nevertheless, falsification still ran into a number of problems. Apart from the issue of demonstrating that one moves closer to "the truth" by falsifying more and more theories, there also exists the problem of how one knows when a theory has been correctly falsified; Popper's own caveat concerning the theory-dependence of observation admits no direct or unfettered correspondence with reality. Popper responded to these criticisms by suggesting that "rational criticism" would allow us to know that one theory was closer to the truth than another, but he was unable to elaborate on how it could do so other than by pointing to essentially educated guesses justified by so-called rational criticism. Barry Hindess notes that, according to Popper,

> "Rational criticism" is to be used to provide "a good critical reason in favour" of the guess that one theory really is closer to the truth than a competing theory. Thus, the only argument that Popper can offer to support the

assertion that "rational criticism" does lead to the growth of knowledge is itself dependent on the method of "rational criticism." In effect we are asked to accept that rational criticism leads to the growth of knowledge because rational criticism gives a good reason for accepting that it does.[18]

Lakatos and Kuhn were acutely aware of the problems and ambiguities surrounding Popper's description of how falsification worked in practice. In particular, they saw minimal resemblance between the process of falsification—which Popper had reasoned to be the only rational form of scientific method—and the actual development of many landmark theories and discoveries. Indeed, as Lakatos and Kuhn both pointed out, a retrospective application of falsification would render the theories of nearly everyone from Copernicus to Newton—and more recently Einstein as well—either falsified or untestable at some point, and thus all such work should have been discarded as either disproven or unscientific. Kuhn described the basic problem facing falsificationists as follows:

> As has been repeatedly emphasised before, no theory ever solves all the puzzles with which it is confronted at a given time; nor are the solutions already achieved often perfect. On the contrary, it is just the incompleteness and imperfection of the existing data-theory fit that, at any time, define many of the puzzles that characterise normal science. If any and every failure to fit were ground for theory rejection, all theories ought to be rejected at all times. On the other hand, if only severe failure to fit justifies theory rejection, then the Popperians will require some criterion of "improbability" or of "degree of falsification." In developing one they will almost certainly encounter the same network of difficulties that has haunted the advocates of the various probabilistic verification theories.[19]

The release of *The Structure of Scientific Revolutions* in the early 1960s had a profound impact on the philosophy of science and scientific thinking for two important reasons. Kuhn's approach to explaining scientific method broke with the traditional approach of defining it in logical and rationalist terms (as Popper had attempted) by instead choosing to explain science by looking at its history. And, more importantly perhaps, Kuhn's historical treatment of science succeeded in severely undermining falsificationism as a means of distinguishing between science and pseudoscience by proposing that no such distinction could be made.

For Kuhn, science has been a mostly uncritical sequence of attempts at "puzzle solving" contained within a progression of paradigms that rise and fall as the problems generated by each paradigm reach a crisis, requiring a new paradigm to solve them and thereby dispensing with old modes of thinking and making way for the new. Scientists, for the most part, have been unquestioning of their methods and have remained locked within the dominant paradigm until a sudden "gestalt switch" in their perceptions provides for a new way to approach currently unsolvable problems. Thus, there is nothing logical or reasoned about how one theory replaces another. It is not the logic of argument that moves scientists from one paradigm to another but only the weight of numbers of scientists who have been converted by the gestalt switch in perception.[20]

Lakatos, however, remained interested in further developing Popper's ideas rather than rejecting them completely as was the case with Kuhn, who saw no grounds for belief in a science based on any universal form of rationalism. Lakatos's ideas differed from Kuhn's in that Lakatos believed a common thread or methodology in scientific discovery could be found that would stand up to the "tribunal of history." At the same time, he accepted Kuhn's argument that the Popperian notion that this methodology could be demonstrated a priori and independently of historical evidence was flawed.

But while Kuhn and Lakatos agreed on the problems in Popper's analysis and description, they had little in common when it came to providing their own ideas on demarcation. Where Kuhn saw scientific discovery as a largely uncritical endeavor trapped within the confines of the prevailing dominant paradigm, Lakatos disagreed because he could not accept the political implications of Kuhn's ideas. The proposition that scientific discovery has been based upon the manipulation of history and unchallenged orthodoxies was too Orwellian for Lakatos and did not sit well with his Hegelian background, which required belief in the existence of some kind of rationality and sense of progress.[21] For Lakatos, the challenge was to develop a theory of scientific method that was consistent with the history of science but that was at the same time rationalistic enough to avoid the relativism of Kuhn's paradigms. Lakatos's response was to dispense with Popper's falsification criterion and to develop instead a methodology based on the internal strength of individual research programs.

For Lakatos, the important feature of good science was not whether a theory could be (or had been) falsified; rather good science was marked by the heuristic or problem-solving methods at the heart of research that drove the program forward by producing novel and unexpected problems and answers. According to Lakatos, a research program—due to its dialectical nature (i.e., the Hegelian dialectic—overcoming the contradiction between thesis and antithesis by means of synthesis)—could continue even if its theories were falsified, provided the heuristic showed promise that such inconsistencies could be explained and accounted for as the research progressed.[22]

In response to Popper's question of when a theory should be abandoned, an important demarcation issue, Lakatos argued that the criterion should depend on the problem-solving ability of the heuristic. In other words, a theory need only be discarded if the research program ran into problems its heuristic was unable to solve and that were preventing it from being both theoretically progressive (i.e., modifications were leading to unexpected predictions) and empirically progressive (i.e., some of its predictions were being corroborated empirically).

But this approach led to the same fundamental problems of epistemology encountered by Popper when trying to articulate falsification criteria and make judgments about one theory being better than another. Lakatos's views on when and how a research program should be labeled "progressive" remained unclear and seemed to allow for the justification of pursuing virtually any research program, since it is only possible to judge a program's failure in retrospect.[23] Thus, at the time of doing the research, it is impossible to know when one should abandon a given program. This flaw in Lakatos's ideas, therefore, meant that no demarcation between science and nonscience is available to those who may be wondering if what they are doing is scientific or not. And this problem was compounded by the fact that even with the aid of retrospect, Lakatos remained unable to explain how a theory could be regarded as better without some recourse to its ability to explain and describe "reality."

Neither Popper nor Lakatos was able to explain how the replacement of one theory by another could be justified without resorting to a form of essentially empiricist method. So their attempts at locating a universal scientific method and, therefore, a means of distinguishing science from pseudoscience ran into trouble when trying to account for progression in science. Kuhn, on the other hand, avoided these problems by accepting the political and value-ridden nature of science, concluding that no universal method exists. Kuhn, however, was uncomfortable with the relativist argument he had created (and the ensuing antirelativist criticisms he attracted); he resorted to trying, rather unconvincingly, to relocate his view of science in a nonrelativist framework by reiterating his belief in "scientific progress" and claiming that one theory could be better at puzzle solving than another.[24] But Kuhn was unable to explain why this would be so without hinting at some return to the problem of a theory's ontological basis mirroring or at least resembling the "real world"—the same problem that vexed Popper and Lakatos in their attempts at explaining scientific progress.

Thus, it appears that the quest for a universal and rational scientific method remains as elusive as ever, raising the important question of whether one actually even exists to be found. Kuhn's seemingly unintentional relativist portrayal of scientific development generally remains the most convincing idea, given the inherent problems of the rationalist and empiricist attempts at justifying direct correspondence between the real world and the realm of thoughts and concepts. The desire to identify some means by which the truth can be sorted from the chaff is easy to understand, as a world made up of only relative truths and theoretical conjecture is bound to make many people uncomfortable, particularly when they board aircraft or undergo medical treatment.

Seeing science as less than "the truth" about what we assume is an independent reality and instead as just another mode of discourse—no more or less capable of adjudicating on the nature of reality (including the nature of society and politics) than any other mode—is a difficult pill to swallow. But, as Ziauddin Sardar has observed, it is a pill we need to swallow if we are to accept that science is more likely about uncertainty and competing opinions than it is about certainty and truth:

> Science is simply not what realists and idealists claim it to be. Its ideological and value-laden character has been exposed beyond doubt. But it is not simply a question of how political realities of power, sources of funding, the choice of problems, the criteria through which problems are chosen, as well as prejudice and value systems, influence even the "purest" science. . . . It is more an issue of how science is now associated with uncertainties and risk.[25]

Still, as Russell explained, having described causality and induction as "the great scandals in the philosophy of science since the time of Hume," the majority of people in modern societies want to continue believing in science even though Hume "made it appear that our belief is a blind faith for which no rational ground can be assigned."[26] The reason is a relatively simple one, and it is eagerly embraced in societies where liberal values are the norm: science gives us control over our environment and therefore our lives. The flip side of this reasoning, however, is that if we accept science as truth, we also must accept that our ability to control our lives is limited by the laws upon which science relies for its rationality:

> Science as it exists at present is partly agreeable, partly disagreeable. It is agreeable through the power it gives us of manipulating our environment, and to a small but important minority it is agreeable because it affords intellectual satisfactions. It is disagreeable because, however we may seek to disguise the fact, it assumes a determinism which involves, theoretically, the power of predicting human actions; in this respect it seems to lessen human power. Naturally people wish to keep the pleasant aspect of science without the unpleasant aspect; but so far the attempts to do so have broken down. If we emphasize the fact that our belief in causality and induction is irrational, we must infer that we do not know science to be true, and that it may at any moment cease to give us the control over the environment for the sake of which we like it. This alternative, however, is purely theoretical; it is not one that a modern man can adopt in practice.[27]

Russell is suggesting that without science as a means of determining what is true from what is not, our ability to control our environment, and therefore our choices, remains limited. Indeed, Popper and Lakatos (and others) rejected Kuhn's ideas because of the implications his relativist views had for any description of how human decision making might be defined. If our decisions are not based on "the truth," or at least on scientific knowledge that is guiding us closer to the truth, the whole notion of human progress becomes questionable. Thus, their objections (and those of others who oppose relativist explanations of epistemological problems) were based more on views of what the world and humanity's role in it should be rather than on any convincing arguments that the world is not in fact "a relativist nightmare," at least as far as epistemology is concerned.

At the risk of resurrecting a hoary old relativist paradox, it seems the only conclusion concerning the supposed direct correspondence between knowledge and an assumed reality that can be argued rationally is that no such correspondence can be logically demonstrated. And with this assumption it still may be possible to establish a sense of human progress and development based on a mode of knowledge creation that is clearly distinguishable from all others.

The obvious alternative to demanding or expecting certainty and truth from science is to accept the conclusions of what our reasoning tells us science can and cannot do and then to redefine our expectations of it in those terms. This of course leads to the question of how one would choose one theory or explanation over another without the criterion of truth as a measure. The most likely answer is to employ a concept already in use but seldom recognized as important in determining what scientific research is encouraged and the ends to which such research is pursued: the concept of utility.

SOME ALTERNATIVE CRITERIA FOR JUDGING SCIENCE

It is perhaps not surprising that our general reluctance to let go of our beliefs in truths, universal laws, and constants—and also in our ability to control and manipulate our environment—causes many of us to recoil from uncertainty and to insist upon certainty in our lives, even if there is no good reason for believing it is available. The extent to which science and technology have improved many human lives over the last two hundred years has led most people to equate science with certainty and truth in spite of the many arguments to the contrary.

But demands for certainty from science are not uniformly spread, and our tolerance for uncertainty appears to be determined by modern society's perceptions of the related costs and benefits, in addition to beliefs in the likelihood and potential effects of possible outcomes. The levels of scientific uncertainty in modern medicine, for example, remain deceptively high but this does not prevent most people from visiting doctors when they are sick, because they believe doctors know what they are doing—even if they don't, the possibility of likely benefits (i.e., getting better) is more attractive than the alternative of remaining sick.

Sciences such as medicine, physics, and chemistry are generally held in high regard due to their perceived utility for society in general. This standard—the utility factor—I call "criterion I." And it is the weight and impact of their benefits that allow the numerous failures and mistakes in these fields to be quickly and easily forgotten. Why? Because many of the results of these sciences appear tangible and are recognizable as having a positive effect (in most cases, though not always) on our lives. Medicine is seen to be progressive due to the increasing number of ailments doctors seem able to treat successfully; physics provides the proofs of its progress in the form of space missions, bigger and faster jets, and other previously unimagined feats of engineering; while chemistry creates wonder drugs, miracle glues, and useful new bathroom cleaning solvents. Even meteorology somehow manages to maintain the mantle of science with occasional success in its ability to predict the "reality" of the weather.

The point here is that these disciplines should not be seen as "progressive" in the sense that Popper or Lakatos prescribed (i.e., moving closer to the truth about reality), but rather by the extent to which they provide recognizable utility to the societies that support and encourage them. The definition of utility I am employing here is similar to that proposed by Jeremy Bentham whose "principle of utility" relies on a fundamental distinction between pleasure and pain, good and bad, desirable or undesirable, and so forth, depending on how these concepts may be defined by the society in question.

> By the principle of utility is meant that principle which approves or disapproves of every action whatsoever, according to the tendency which it appears to have to augment or diminish the happiness of the party whose interest is in question: or, what is the same thing in other words, to promote or to oppose that happiness. I say of every action whatsoever; and therefore not only of every action of a private individual, but of every measure of government.
>
> By utility is meant that property in any object, whereby it tends to produce benefit, advantage, pleasure, good, or happiness, (all this in the present case comes to the same thing) to prevent the happening of mischief, pain, evil, or unhappiness to the party whose interest is considered: if that party be the community in general, then the happiness of the community: if a particular individual, then the happiness of that individual.[28]

Bentham's principle of utility is intended to explain, as an absolute proof, what the basis of human choice ought to be in the moralistic and legislative senses. But in this study I use it only to describe what I contend is being used as the basis of decision making for people, groups, and governments; "utility" as such ought to be recognized as the standard against which forms of knowledge are measured in practice.

Bentham admits that there is no "direct proof" of his principle's veracity but contends, in the vein of the all too familiar circular argument, "that which is used to prove everything else [i.e., the principle of utility], cannot itself be proved: a chain of proofs must have their commencement somewhere. To give such a proof is as impossible as it is needless."[29] Clearly then, Bentham was unconcerned with the problems of knowledge outlined earlier, but this does not mean his utility principle is without application in the descriptive sense I just mentioned.

The concept of utility I am proposing, drawn from Bentham's apparent inference that utility is judged by the party concerned, is essentially a relative one in so far as the notions of happiness, pleasure, pain, and so forth used to define "utility" are dependent upon the established customs and norms of a given society or group of people. For the purpose of explaining why some theories more than others are judged problematic in terms of uncertainty (i.e., are marginalized or rejected), and by some societies but not by others, it is not necessary (nor possible, I believe) to establish any universal standard of what constitutes utility. It is only necessary to provide a logical and coherent argument explaining how and why relative notions of utility have been used, in practice, as the standards against which science and other forms of knowledge are measured.

My intention in providing such an explanation—which is essentially the brief of the following chapters and their discussion of scientific uncertainty in the IWC—is to illustrate but one example (the IWC) of the broader political environment in which scientists necessarily operate, and to explain how that environment judges science on the basis of various perceptions of utility. The implication here is that the notions of "truth" used to judge science by governments and societies are determined by utility, rather than by any absolute standard demonstrably representative of reality. Such notions are therefore variable depending on the priorities or "needs" of the group or society in question—priorities that, in turn, determine the kind of truth being sought (i.e., "all whales are endangered" or "only some whales are endangered"). Thus, by looking at the treatment of science in terms of its perceived utility, rather than simply its verisimilitude in relation to reality, we may be able to more clearly see the political ends to which science is used and directed and the role played by uncertainty in achieving such ends.

The relationship between "utility" and "needs" is important for this argument and requires some elaboration. Put simply, if an individual, government, or group believes that some action or belief about the world fulfills or matches an established or popularly recognized need, that action or belief is then deemed to have utility. If misery, discomfort, and pain are to be avoided and happiness and pleasure preferred, it is reasonable to assume that there are certain essential needs that must be met, such as access to food, clean drinking water, and clothing and shelter. But while it can be argued that the only essential biological needs, depending on one's environment, may be air, food, and water, I would contend that it is not possible to objectively justify only these items as basic needs, nor draw clear distinctions between "wants" and "needs" in any universal sense—given the great differences between the cultures, customs, and modes of production and organization in the world's various societies.

By claiming that I can objectively know the difference between someone's wants (i.e., something they do not necessarily need) and needs, I must also claim to know what that person needs; in this way the task of distinguishing between wants and needs and of establishing absolute needs for others are inextricably linked. The problem here, however, is how another party could rationalize such a judgment on someone's behalf without reference to some privileged form of knowledge or objectivity.

To make judgments about what one person needs as opposed to wants is highly problematic, since such judgments assume the right to hold influence over the way others lead their lives based on what is supposedly best for them. This raises, among other things, the controversial issue of paternalism, to which liberal utilitarians like Mill have taken great exception.[30] Put another way, if someone does not want something (or cannot want something because he or she does not know it exists), in what sense can it be said that they nevertheless "need it" without assuming to know what is best for that person? And on what epistemological basis could it be assumed that one did in fact know without encountering the problems outlined earlier? Indeed, as Len Doyal and Ian Gough have pointed out, any such distinction "tells us more about those who make it than it does about the human condition."[31]

The principle of utility might be used to justify such a position but only on the basis of individual choice and established cultural norms, not on some objective standard of what is or is not needed versus merely wanted. In these terms, it seems reasonable to assume that any distinction between needs and wants cannot be judged in absolute or objective terms and should, therefore, be a relative one based upon a given set of norms, customs, and beliefs.

It may, however, be possible to determine the needs of others in a practical and transcultural sense while recognizing that any such determination must necessarily be subjective or relative (I do not intend to argue that one can never help someone by giving that person something one believes is needed to make the person happier and more comfortable). Doyal and Gough have recognized the problems mentioned so far but nevertheless claim that a definition of "objective and universal human needs" is possible.³² They propose that "health and autonomy are the basic needs which humans must satisfy in order to avoid the serious harm of fundamentally impaired participation in their form of life."³³

Doyal and Gough's argument is compelling because it is based on the contemporary liberal standards concerning what is or is not morally or ethically defensible that guide what most of us in so-called modern and civilized societies believe to be reasonable behavior. But, like Popper and Lakatos, their theory ultimately depends on notions of what is or is not politically desirable, which cannot be demonstrated in any absolute or objective sense. However, in the following passage, Doyal and Gough, in seeking to explain the compulsion people often feel to help others and the failure sometimes to do so, do offer what could be considered a practical and rational gauge for determining the needs of others and our own responsibility toward providing such needs, without requiring a claim to objective knowledge:

> Thus, to see a starving mother in another society trying unsuccessfully to feed her infant carries with it the conclusion that she should be better able to fulfil whatever obligations toward her child we impute to her. Given such imputations, we take on the obligation to try *in some way* to help. The general response that "something should be done" to satisfy the minimal needs of those in even radically different cultures is due in part to the fact that they hold up a nightmarish mirror in which those in more comfortable nations can see themselves reflected—knowing that they should act in some way but finding themselves too incapacitated to do so. To use Shakespeare's image, they see "the heath" and themselves as the "poor, bare, forked animal" unprotected on it.³⁴

Here, the solution to justifying a judgment comes through a simple recognition of how we define our own happiness and comfort, rather than from a universal standard, and the realization that there is no rational basis for expecting such happiness for ourselves while denying it to others. In effect, suffering among others—which on the most fundamental level we attribute to deprivation of basic needs—reminds us that we could experience similar circumstances at some point: in which case we would hope for and expect the help and understanding of others.

Basic needs can be defined, then, in a general sense, given common standards of suffering that people generally hope to avoid. But beyond this, definitions of needs in even a general sense become highly problematic and can only be referred to in relation to the customs and norms of a given society. The necessarily subjective nature of these needs, I contend, determines the notion of utility and how it is used as a standard against which knowledge is measured, given the absence of any direct correspondence with reality—a single reality whose existence can only be assumed—and also given any objective way of describing reality that might otherwise provide an objective standard of knowledge demarcation.

The distinctions people draw between different forms of knowledge, therefore—in addition to the justifications used to separate genuine knowledge from simple beliefs—are essentially based on value judgments relating to need-satisfaction (i.e., what I have described as "utility"), rather than on any criterion capable of comparing a theory's claims and predictions with reality in a direct and unproblematic way. The conviction that the physical sciences are capable of revealing and explaining reality has largely survived in the public eye, in spite of the arguments against such a proposition, but only because of their ability to continue providing tangible benefits that people see as satisfying recognized needs. Knowledge demarcation, then, is not based on a scientific method capable of accurately describing and predicting our physical world, but is instead grounded in value judgments about what people need and the extent to which those needs are satisfied. Put differently, scientific methods are judged by the utility they are seen to provide, and not by their ability to describe reality, since there is no means of directly corresponding with reality and, therefore, no way of verifying that a theory or method is able to do so.

But in contrast to the physical sciences, the social sciences—sometimes regarded as the poor cousins of the family and some of which Popper targeted in his quest to distinguish between science and nonscience—cannot provide the same kinds of tangible benefits or need-satisfaction to society because of their very subject matter: societies and the behavior of people both past and present. The nature of these subjects is empirically nebulous and often difficult to define. Thus, political science, history, economics, anthropology, and psychology are all relegated to some level of esteem below science simply because of their inability to provide benefits that people can use, listen to, ride in, or watch in their daily lives.

However, of all the so-called social sciences, history—as it is popularly (mis)understood (i.e., a discipline that can reliably inform us of the who, what, when, why, and how of particular events)—appears the closest to the physical sciences in terms of satisfying criterion I (perceived utility). This partly results from the more accessible nature, generally speaking, of history's empirical method and content (documents, observations, and interviews) but, more importantly, it also comes from the significant influence history has on people's interpretation and understanding of events and the utility historical analysis provides in justifying current attitudes and thinking. Like all forms of knowledge, however, the study of history is confronted by the same epistemological problems facing the natural sciences; these issues have divided and frustrated many historians for most of the postwar era, although, as in other disciplines, not all historians have accepted the consequences these problems entail.[35]

But for most people, and also for the requirements of criterion I, it is unimportant that the reasons scientists give for airplanes flying or for people dying of cancer may be shown to be completely wrong in another hundred years. People do not accept current explanations for what happens around them because they *know* them to be true, but rather because they *believe* them to be true as a result of the utility these explanations are seen to provide. This attitude explains, for basically the same reasons, why many Americans, Australians, Britons, and others largely are indifferent to the possibility that the currently accepted rationale offered by many, but not all, historians for the U.S. government obliterating Hiroshima and Nagasaki in 1945 may turn out to be wrong. After all, the utility of considering that hundreds of thousands of people may have been killed unnecessarily, or that we may not know exactly why 747s usually stay in the air but sometimes crash, is no doubt highly questionable for people who believe they must know that the use of atomic weapons against Japan was justified or that flying is based on proven aerodynamic principles.

What does matter is that what some scientists (or historians) say today usually seems to correspond better and more comfortably with their lives and experiences than what other scientists were (believed to be) saying one hundred years ago. And it is in this sense that science is naïvely assumed to be progressing in an orderly and rational manner and, therefore, moving steadily toward the discovery of more and more truths about the world. Such progress, however, is difficult to find in the social sciences. The bottom line then, in determining acceptable levels of scientific certainty, appears to be the issue of utility and the extent to which a given field of endeavor can provide enough of it to satisfy people's needs and aspirations—however they may be concocted by that thing we vaguely define as society.

Thus, people's demands for certainty are tempered by the extent to which their needs are satisfied (criterion I). But a second criterion—"criterion II"—

also needs to be met if serious opposition based on uncertainty is to be avoided: the extent to which satisfying those needs outweighs any perceived risks or disadvantages. Any such disadvantages can be characterized as the compromising of another already established and presumably satisfied need. An example of this dichotomy between cost and benefit is the extent to which people in industrial and postindustrial societies are prepared to tolerate global warming's effects on the environment in order to maintain current modes of transport, power generation, and economic productivity.

The first criterion (criterion I) of satisfying society's so-called needs for faster and more comfortable transport, more effective medicine, and cleaner bathrooms has so far been sufficient to largely disqualify the social sciences from achieving any notion of tangible progress and, therefore, any significant relevance to the reality of people's lives in general. But while the physical sciences are, in the naïve sense at least, able to meet this standard, it is important to note that this standard also determines the kinds of scientific research that are pursued and for whose benefit research is conducted.

One obvious example of this is the great progress pharmaceutical companies made toward helping men in developed countries maintain erections while more than one million men, women, and children in lesser-developed countries continue to perish for the want of a cheap, effective malaria vaccine. Even though scientific research might, with a very high degree of certainty, dramatically reduce the death rate caused by malaria, such research is yet to happen primarily because the funding and expertise needed can only come from developed societies.[36] Unfortunately, these societies place little or no priority on preventing malaria deaths because such attention is neither profitable nor part of the perceived reality of their existence.

As another example, would modern medicine be any closer to a more effective treatment for the horrifying and almost always fatal Ebola virus if outbreaks of the disease regularly occurred in North America or Europe rather than only in Africa? Dr. Timothy Stamps, Zimbabwe's former Minister for Health and Child-Welfare, noted in 1998,

> there is virtually no investment in anti-malaria drug-development in the commercial sector of the pharmaceutical ethical organisations, probably less than $30 million every year.... Of course the group which will benefit from new drugs is the poor, and they cannot afford to buy them so there

is very little point if your philosophy is to look after your shareholders rather than the stakeholders. There is very little point in developing a drug for the poor! Dr. Wellcome must be turning in his grave.[37]

The second criterion (criterion II), which requires that the benefits of a given piece of scientific research not adversely affect the satisfaction of already established (and also highly subjective) needs, is more concerned with implementation. This criterion, however, is often a more difficult one for scientific innovations and research to overcome even if they have the potential to "satisfy a need" in the criterion I sense—as demonstrated by the continuing controversy over the effects, causes, and extent of global warming in spite of the generally recognized need for countries to remain above sea level. The social sciences also experience problems with this criterion, but less frequently due to their inability to satisfy the first criterion in most instances.

Another very good and very topical example of the problems science can encounter when facing the second criterion is the issue of genetically modified food (GMF). GMF scientists have demonstrated to most people's satisfaction their ability to grow more food more cheaply, faster, and more reliably using GMF technology—benefits that clearly satisfy a common need in every society. However, the dangers and risks, real or imagined—that have been most vociferously raised by people in developed societies claiming that these benefits may conflict with other already established needs (i.e., biodiversity, safe food, freedom of choice)—have seriously undermined the veracity of GMF science and its claims that the technology is safe. Although some people (most notably the Prince of Wales) have questioned the ability of GMF scientists to satisfy the common need to grow more food faster and more efficiently, many critics question scientists' ability to do so safely, therefore also questioning the scientific method and reasoning that underpin these claims. Enter the precautionary principle.

LOOK BEFORE YOU LEAP OR HE WHO HESITATES IS LOST: A TALE OF TWO PROVERBS

The responsibility for "proving" that GMF will never create future costs and dangers currently deemed to be unacceptable is being forced upon its creators and supporters without any similar requirement for proof expected from those who oppose the use of GMF—a common criticism of the precautionary principle that will be more fully explored in later chapters. As we have already seen from the problems that have prevented science from proving its endeavors, in either the empirical or rationalist sense, there is certainly no way for scientists or anyone else to prove anything in any universal or timeless way, which is what opponents of GMF appear to be demanding.

Of course, the opponents of GMF are equally unable to prove that GMF is potentially dangerous in the sense of criterion II, but they are able to speculate about the future, essentially on the basis of the unavoidable uncertainty concerning what may or may not happen next. Thus, while scientists argue there is currently no significant evidence to support the contention that GMF will incur costs and dangers, their opponents counter by saying the lack of scientific evidence is not sufficient reason for saying costs will not occur, since the future necessarily implies uncertainty due to the imperfect nature of science's predictive capabilities.

This approach to science by those who believe an innovation or piece of research fails to meet the conditions of criterion II is based almost entirely on judgments concerning established needs and how they may be affected, and this mode has risen to prominence since the 1970s under the loosely defined concept known as the "precautionary principle." As already mentioned, the precautionary principle attempts to deal with the uncertainties science necessarily produces and is made attractive by the simplicity of its "look before you leap" approach to managing scientific unknowns. A brief summation of its various manifestations as an environmental management principle (the area in which it has become most common) describes it as "a principle which proscribes any activity which may cause serious environmental damage even if there is no clear scientific evidence to support the belief that any such damage may occur."[38]

The precautionary principle, then, removes the onus of explaining and justifying doubts from those opposing an activity or piece of scientific research to those wishing to pursue such an activity or research. As a commonsensical notion it is certainly attractive, but its simplicity and lack of clarity raise as many problems and dangers as it is intended to solve. Among these is the absence of any clear guidelines for how levels of acceptable and unacceptable uncertainty can be determined. Also lacking is any definition of the process by which groups or governments invoking the precautionary principle can or should prioritize costs and benefits in determining when and where the principle should be applied.

The invocation of the precautionary principle by governments and nongovernmental organizations (NGOs), as the IWC experience will show, has been determined largely—if not entirely—by the standards of criterion II (the extent to which perceived risks or disadvantages outweigh the benefits of a course of action). The major problem for science presented by criterion II, if one is to avoid the obstacles to implementation posed by the precautionary principle, is the need to avoid falling back on empirical notions of science in order to justify pursuing an activity or research that might damage the interests of someone or something (e.g., whale populations, the stratosphere, etc.).

That said, I do not believe it is either useful or possible to try to make decisions without reference to empirical observation, since it would be hypocritical to deny that we are all forced to do so by the absence of any other means of interpreting and managing what goes on around us. The fundamental problem of epistemology as I understand it—which has foiled the attempts of Popper, Lakatos, Kuhn, and many others and also, I believe, has led to unrealistic demands being made of science in the form of the precautionary principle—is the assumed need to explain and describe the existence of a rational scientific method and to demonstrate that science is somehow progressing toward "the truth."

The expectation of science, then, seems to be that someday, through sheer force of will and endeavor, it finally will provide us with the ability to know everything. But this belief that, to borrow the *X-Files* slogan, "the truth is out there" is only that: a simple belief, with no more rationality or substance to support it than our equally simple desire for it to be so. Therefore, it is a fundamental mistake to explain choices and their related justifications only in terms of how and whether or not we know them to be true or accurate in accordance with scientific method.

It is indeed difficult to imagine how environmental damage could be guarded against without some use of empirical observation methods. Even the unattainable ideal (of some) of avoiding any potentially negative interaction (whatever that might mean) with the natural environment is based on empirical thinking in the sense that empirical observations, colored by perceived human needs (i.e., values), are what inform supporters of this view about what is or is not potentially negative.

So if we accept that we are stuck with empirical methods in the absence of anything else, it seems pointless to dwell on questions about who is or is not justifying their views with empirical observation, since everyone is in some form or another. Consensus among scientists occasionally does occur, but such cases do not represent a removal of uncertainty and are more about "believing" than "knowing." Thus, consensus on an issue is better explained as a situation where one particular interpretation of the empirical evidence becomes convincing enough in the given circumstances for scientists to *believe* it to be more convincing than any of the alternatives, but this is *not* the same as knowing; the emergence of an alternative, and even more compelling, explanation of the data at some future point always remains a possibility.

Thus, because we cannot establish that science can directly describe or correspond with the "real world" and is, therefore, unable to remove uncertainty completely, it also seems pointless to try to explain competing scientific interpretations of reality purely on the basis of which one is more or less true. The main question we should be asking—indeed, the fundamental question I will address—is, "To what ends do governments and NGOs use empirical scientific methods and why?" This should be our focus if we are to make any progress in understanding what scientific advice represents in environmental issues and how scientific uncertainty is used to influence policy.

In the IWC (and in many other examples), empirical methods are employed to justify various points of view, which are themselves derived from empirically based and, therefore, ultimately subjective beliefs concerning what is or is not desirable or useful as per criterion I. So it follows that opposition to whaling is based on past experience of whale hunting, scientific evidence describing the important role of whales in the marine ecosystem, and also scientific research claiming that whales are very intelligent, for example, while those who support whaling do so by arguing about essentially the same issues from a different perspective. Scientific knowledge concerning both (or any) perspectives on these issues is more or less equally limited by scientific uncertainty and thus one is unable to conclusively justify a position by simply resorting to scientific knowledge alone—hence the IWC's ongoing disputes about scientific uncertainty.

Indeed, if environmental debates were only about competition between different scientific theories seeking to describe the world and human effects upon it, then one would expect all debates to end eventually in stalemate, with the differing proponents accepting the likelihood that their theory was no more or less true than anyone else's—since no theory is able to conclusively prove itself in any timeless or absolute way. Environmental (and other) policy debates, however, rarely if ever end this way. Decisions one way or the other are always made at some point, since one view is always chosen over others due to the political and economic necessities that demand decisions be made. But if we accept that these decisions cannot be based on scientific knowledge and theories alone, it then becomes necessary to consider what other factors influence and determine our choices.

My argument contends that our choices are fundamentally determined by their perceived utility. And since the notion of utility I am using is entirely a subjective one, different people, groups, societies, and cultures will perceive the utility of a given choice differently in relation to the values and cultural preferences they subscribe to. The likelihood of uncertainty being invoked, then, in order to oppose a given piece of research or technology—on the basis the precautionary principle—is dependent upon the perceptions of that research and technology that people, governments, and groups hold, as defined by criterion I (the extent to which it fulfills a recognized need) in the first instance, and then by criterion II (the extent to which it is not believed to conflict with an already established need, or the willingness to trade one for another).

When looking at decisions and judgments concerning science in this way, science itself, perhaps, is really only part of the story. Scientific research, presented in these terms, becomes a vehicle that, in the final analysis, only represents the pursuit and justification of political goals, rather than a noble quest for the truth, which many have believed to be the ultimate guide in the steady march of human progress. The perspective I have outlined here may reveal a great deal about why scientists, governments, and groups citing the same methods and data so often fail to arrive at the same conclusions. Such controversies, I suggest, are usually more about what conclusions science should provide, and not simply about earnest endeavor to reveal "the truth," since studying what science represents and how it is used tells us more about people than it does about the natural world.

The actual science involved in the controversy concerning the effects of whale hunting, therefore, probably only partially explains what is going on in the IWC in much the same way that knowledge about the effects of smoking only partially explains why people choose either to smoke or not to smoke. With smoking, as with any issue or problem, one can adopt one of four basic positions concerning its effects: (a) I believe smoking is harmful; (b) I believe smoking is not harmful; (c) I believe smoking can be both harmful and beneficial; or (d) I have no idea. People choose to smoke or not smoke partially because of the position they adopt in relation to the knowledge they choose to believe but mostly, I contend, because of the utility they believe smoking or not smoking provides—utility they recognize through experience and explain in terms of knowledge.

Most smokers no doubt believe their habit is not good for them in the physiological sense, but they continue to smoke because of the utility they believe smoking provides. Conversely, do nonsmokers reject smoking purely because of the associated health risk or also because of the lack of utility they see in smoking? If nonsmokers were to accept that smoking may relieve stress or aid weight loss, as some medical research has suggested,[39] would those suffering from chronic stress or obesity then perceive enough utility in smoking to become smokers, despite believing that it is bad for them in other ways? Similarly, if anti-whaling advocates were to suddenly discover some utility in whaling, would it be unreasonable to expect at least some to change not only their position on whaling but also their perception of scientific uncertainty concerning the dangers of whaling?

TO BE CONTINUED: A HYPOTHETICAL ILLUSTRATION

For purely illustrative purposes, consider the following hypothetical situation that encapsulates most of the major issues and problems outlined thus far. This scenario broadly summarizes the major issues my analysis sets out to address and also provides an easily remembered touchstone with which readers can follow the arguments and problems I pursue in the following chapters. This entirely hypothetical scenario, however, will remain open-ended until the book's conclusion, where we will again consider the questions posed, at that point with a detailed analysis of the IWC's experience with scientific uncertainty in mind.

Let us suppose that, at some point in the very near future, medical science produces a cure for cancer that passes rigorous empirical testing and numerous falsification attempts. The new drug is widely hailed as a turning point in modern medicine and is predicted to save millions of lives. As a result, it easily meets the standards of criterion I, and public demand for its usage swells to almost irresistible proportions. However, an animal rights group reveals that commercial production of the drug requires large amounts of oil from the minke whale—a species the IWC Scientific Committee considers to be unendangered but one that the commission continues to protect from commercial hunting.

Suddenly, the societies of the IWC's anti-whaling governments are embroiled in controversy surrounding the new cancer cure, as its release is now complicated by the imperatives of criterion II. Various environmental and animal rights NGOs are thrown into turmoil over their priorities and the threat of alienating the members they rely upon for funding, while the news media are awash with views and opinions about the protection of whales and the sanctity of human life. Governments are also in a state of confusion and delay any decision on the drug's release in order to review further the scientific evidence and advice on both the drug and the sustainability of limited minke whale hunts.

As a consequence, the precautionary principle is brought under heavy scrutiny, as the basic policy choices of continuing to "look before you leap" and picking up the gauntlet of "he who hesitates is lost" stare government leaders and their advisors squarely in the face. What policy will the various governments adopt concerning the manufacture and release of the drug? And how, if at all, will this dilemma affect each government's position on scientific uncertainty concerning the dangers of managing a return to commercial hunting?

2

THE IWC 1949–59

AN EXERCISE IN UNCERTAINTY BECOMING CERTAINTY

Under the IWC's stewardship during the 1950s, hunting was mainly focused on three species—the blue, fin, and humpback. Each of these species, despite warnings from the commission's Scientific Committee, was hunted to the point of near extinction in the Antarctic, where the majority of whaling operations occurred.[1] In retrospect, it is tempting simply to blame the period's excessive hunting on the five governments that effectively controlled Antarctic hunting—the United Kingdom, Norway, the Netherlands, Japan, and the Soviet Union—by explaining their actions only in terms of a single-minded pursuit of profits that gave short shrift to conservation initiatives. But while such thinking no doubt was prevalent in the IWC at the time, and contributed greatly to the depletion of some Antarctic stocks, this account of the reasons behind excessive hunting provides only a partial explanation, since it ignores how perceptions of scientific uncertainty affected management policies during this period.

In the course of explaining the controversy and disagreement that surrounded the IWC's setting of catch quotas and their implementation, this chapter will focus on the treatment and interpretation of scientific uncertainty by the commission's members and scientists. The reasons why the IWC's members continued to approve quotas that were in excess of what the SC believed to be sustainable—the strong postwar demand for fats, the financial pressures caused by intense competition and declining stocks, and the compelling need of various industries to recoup the high levels of investment spent on infrastructure—will be outlined, but only briefly, since the economic forces that drove whaling during this period have already been examined in detail elsewhere.[2]

For the purposes of this study, explanations of why governments and whaling industry representatives argued in favor of what many believed, or at least suspected, to be unsustainable policies are of course important. But of particular importance is the question of how this situation was played out and rationalized in the IWC by the parties involved. The industry's pressing economic imperatives at the time encouraged varying degrees of unwillingness within the commission to accept scientific advice aimed at conservation, but they tell us little about how scientific advice was actually relegated to near irrelevance in the IWC, an organization that had declared its policies "shall be based on scientific findings."

To borrow the parlance of a police investigation, the motive appears to be clear (financial pressures) and the victim easily identifiable (conservation of whale stocks). I will contend, however, that the weapon involved was none other than the invocation of scientific uncertainty. Further, as we will see, this is most often the weapon of choice for governments and organizations seeking to discredit scientific advice that advocates policies that do not match existing political and economic priorities as defined by criterion I (utility) and criterion II (compatibility with already established needs).[3] To illustrate how scientific uncertainty can be employed to pursue a desired policy outcome, we turn to the fin whale debate, which occurred in the IWC during the mid to late 1950s and also shares some important similarities with many of the problems the commission struggles with today.

WHALING BEFORE THE IWC

As many observers and critics have noted, conservation issues were never taken seriously by the international whaling community prior to the Second World War—in spite of several earlier whaling conventions being initiated along with some steps taken (primarily by the United Kingdom and Norway) to curb the excessive hunting that had characterized pelagic whaling for most of its history.[4] In the modern era, heavy capitalization of the various whaling industries in the years following the First World War led to intensive hunting, which resulted in oversupply in the whale oil market by the early 1930s, a subsequent drop in market prices, and further depletion of species such as the blue, humpback, and right whales. Industry attempts to reduce the number of animals taken at this time were at best only partially effective, since these efforts were based more on economic concerns than on any genuine awareness of the pressing conservation issues that would have such a profound effect upon whaling[5]—the scenario, in effect, for the events of the late 1940s and 1950s.

In retrospect, it is perhaps ironic that the international situation accompanying the end of the Second World War heralded both an excellent opportunity to reform the whaling industry and the beginning of a period noted primarily for some of whaling's worst excesses. Indeed, with much of the world's whaling fleets destroyed during the war, the financial imperatives for larger hunts—created by previous overcapitalization of the industry—had largely disappeared along with the many sunken or converted whale catchers and factory ships.[6] A fresh start would have been possible in terms of balancing the economic needs of the whalers with the biological and reproductive capabilities of the whales. Furthermore, the war years had provided a short respite for the stocks and had allowed for some recovery in numbers, although it is now generally agreed that the period of reduced hunting during the war was too short to have had any significant effect.[7]

In the postwar years, however, it soon became obvious that the goals of conserving stocks and using scientific advice to determine quotas, as set out in the 1946 International Convention for the Regulation of Whaling (ICRW), were being ignored for the most part by the IWC's members. The same short-term economic priorities that had so far driven the capital-intensive era of modern whaling quickly reemerged after the Second World War—signaling an ominous and unmistakable return to business as usual for the whaling industry.

Throughout most of its history, the whaling industry has continued its habit of sequentially depleting various stocks: the practice of hunting the most desirable species until sufficient numbers could no longer be found before then moving on to the next most desirable species, and so on. This methodology began with the slow-swimming right whales, which were heavily hunted in the North Atlantic until depletion of these stocks near the end of the 1700s led the whaling fleets to stocks in the Southern Hemisphere and North Pacific. These stocks too had collapsed by the mid-1800s and sperm whaling was also coming to an end about this time.[8]

The introduction of the harpoon gun and exploding harpoon, invented by the Norwegian whaler Svend Foyn in the 1860s, and its use aboard steamships was a revolution that revived whaling by allowing hunting of the faster swimming and previously uncatchable rorqual species such as the blue, fin, and sei. With rowboats and hand harpoons, whalers had been limited to the slower humpback, grey, and right whales. The arrival of steam-driven ships armed with harpoon guns meant that no species was beyond the reach of the whaling fleets. Thus the blue, fin, and sei whales in the Northern Hemisphere, thanks to Norwegian technology, became the industry's main targets.

The era of modern whaling began off Finnmark, a Norwegian county, in the 1860s and progressively spread around the world until the discovery of large numbers of blue and fin whales (and also largely untouched stocks of right whales) in the Antarctic. After 1904, the year of the first southern expeditions, the Antarctic became the main focus of the world's whaling fleets, and Antarctic stocks, in particular the blue and fin, would suffer accordingly over the next sixty years until they too, like the right whales before them, became too scarce to be commercially viable.[9] The final act in this sequence of depletion was played out in the 1970s with Japanese and Soviet hunts of the minke, the smallest baleen whale. But by this time the political environment of the IWC had changed sufficiently—in response to an obviously failing whaling industry—to spare this species the same levels of overhunting experienced earlier by its larger rorqual brethren.

Thus, improved technology, leading to Antarctic whaling, and also the near annihilation of the world's right whale stocks by the early 1930s[10] ensured that the two largest rorquals, the blue and fin, would become and remain the whales of preference for most of the modern whaling period. As a result, much of the IWC's early scientific deliberations and attempts at regulation involved these two species and in particular the Antarctic stocks, since this was where the vast majority of whaling occurred.

OLD HABITS DIE HARD: THE WHALING OLYMPIC

The short explanation of why the IWC almost allowed several species of whale, and itself, to be consigned to oblivion within the first twenty years of its existence is a relatively simple one: irresponsible management based on greed. The price of whale oil in the postwar years almost tripled, increasing from £40 per ton in 1945 to £110 by 1948, and maintained an unprecedented average high of £100 per ton between the years 1946 and 1952.[11] The rapid postwar increase in price was caused by the world fats shortage precipitated by war-related disruptions, a widespread failure of grain and other crops in 1947, increasing populations, and also a shortage of U.S. currency in Western Europe that made the purchase of fats difficult in a U.S. dollar dominated market.[12] The net result of the increased demand for whale oil and its ensuing higher prices was a highly competitive scramble by whaling companies for as large a share as possible of the 16,000 Blue Whale Units (BWU)[13] that had been set as the Antarctic quota at the 1944 and 1946 whaling conferences and was later reconfirmed at the IWC's inaugural 1949 meeting.[14]

Postwar whaling, which began in the 1945–46 Antarctic season, was largely a free-for-all involving anyone able to put together the ships and crews required for an expedition. As we will shortly see, the number of participants was limited by the political circumstances of the time, but such impediments were not enough to prevent the wholesale slaughter that would come to characterize this period.

During the Antarctic season, the various fleets were required to radio in their catch (which could not be independently verified) to the Bureau of International Whaling Statistics in Sandefjord, Norway, each week. When the whalers came close to filling the quota, the bureau then would announce the closing date for the season. The effect of this system was to create what whalers dubbed "the whaling olympic": the fleets killed as many whales as possible in the shortest possible time. In order to catch as big a share as possible, companies would invest in bigger and faster catcher vessels. But increasing investment in bigger and faster catcher boats by all the competitors soon led to a situation where the goal of catching more whales than anyone else was reduced to the hope of only maintaining one's share of the quota, in spite of the voluminous levels of capital the industry was fast absorbing. According to George L. Small:

> The Antarctic quota put every floating factory in a race against time with all other expeditions.... Financial success could be had only by killing as many as possible as quickly as possible before the order to stop whaling came out from Sandefjord. Factory ships and catchers alike worked twenty-four hours a day, seven days a week, weather and whales permitting, until the season was over. Pelagic whaling in the Antarctic was so exhausting and hectic that the whaling men aptly dubbed it "The Whaling Olympic." ...
>
> When a whaling company acquired better catcher boats it automatically acquired an advantage over its competitors. The competitors in turn had to acquire better catchers in order to overcome the disadvantage. Better catchers lead to fewer whales and fewer whales lead to a need for better catchers....
>
> Better catchers did not bring increased production or revenue; they

brought only the hope that each company would catch its proportionate share of a decreasing number of whales.[15]

Thus, the pressures of competition led to a vicious circle of more and more capital being required to catch a steadily dwindling number of whales faster than anyone else. During the boom period, it was possible for profit margins to be maintained, but when the prices began to drop after the 1951–52 season, profitability became increasingly elusive in the face of increasing capital investment. The major problem facing companies was to make enough to recoup the heavy investment that participation in the whaling olympic already had required, which in turn led to increasing reluctance among IWC members to reduce the quota. This situation was precisely what the Norwegian and British governments had hoped to avoid by preventing other countries from rejoining pelagic whaling in the Antarctic. However, while their plans to create a monopoly over Antarctic whaling did effectively limit the number of nations involved, they were unable to exclude everyone.

The defeat of Germany and Japan had left the United Kingdom and Norway as the dominant pelagic whaling nations in the aftermath of the war. The United Kingdom and Norway had been instrumental in initiating earlier attempts at international management and conservation of whale stocks in the 1930s and were the only two countries from which whaling companies were able to continue pelagic hunting throughout the war.[16] Japan and Germany, which together with the United Kingdom and Norway had represented the major whaling nations in the 1930s, were prevented from returning to any significant role in whaling during the initial postwar years by the occupying Allied powers. But while Germany never managed a return to whaling, Japanese whalers were able to recommence Antarctic whaling as early as 1946 under U.S. control and found themselves free to continue whaling independently after the signing of the 1951 peace treaty in San Francisco and its subsequent ratification in 1952.[17]

Norway and the United Kingdom, however, were unable to continue their prewar lead in the shaping of international regulation entirely on their own terms after 1945, since the outcome of the war had given the U.S. government an influential international role and, therefore, also a strong voice in determining the future shape of international whaling regulation in spite its own relatively small interests in the industry. Thus, the task of determining how postwar whaling would be regulated was left largely in the hands of these three governments.[18] At the 1944 whaling conference in London, and again at a second conference the following year, the creation of a standing commission was discussed. Both of these conferences were held primarily to amend the earlier initiatives of the 1937 and 1938 London agreements and to create a regulatory regime for postwar Antarctic whaling.[19] The most significant meeting of this time, however, was the 1946 Washington Conference, which the United States had announced it would sponsor and host at the 1944 London Conference. It was the Washington Conference that produced the International Convention for the Regulation of Whaling and its administering organization, the International Whaling Commission.[20]

But while the United Kingdom, Norway, and the United States were largely responsible for the IWC's creation and the duties it would perform, the governments of the Soviet Union and, in particular, the Netherlands were also developing a strong interest in pelagic whaling and, subsequently, its regulation. The Netherlands wasted little time in establishing its industry as a force to be reckoned with in Antarctic whaling by quickly putting together a whaling operation in 1946 with government support to meet the acute fats shortage being experienced at home. Indeed, the Netherlands' entry into the whaling club would later prove to be a turning point, as the influence exerted by the Dutch government on the international regulation of whaling over the next fifteen years significantly contributed to the IWC's failure to conserve stocks.

Norway and the United Kingdom's governments were both very keen to limit the number of whaling fleets going after the 16,000 BWU quota set at the Washington Conference, since there was already a strong suspicion among some scientists and industry managers that stocks had not recovered to any significant extent during the war—as indicated by the poor catches of the 1945–46 season—in addition to a somewhat prophetic belief that increasing competition would rapidly reduce stocks and bring about the industry's demise. Another, and perhaps even more compelling reason, was the Norwegian and British assumption that new competitors also would make whaling unprofitable in the long run since the boom in whale oil prices was certain to be short-lived. Thus, the Norwegian and British governments strongly opposed German and Japanese fleets returning to whaling and also the Netherlands' bid to develop its own industry. In December 1945, less than a month after the Netherlands' admission to the 1945 London Conference, the Norwegian government even went so far as to impose the Norwegian Crew Law, which effectively banned Norwegian crews from working under any foreign flag other than the

United Kingdom's—although the Soviet Union was allowed crews for the 1946–47 season. This law, in combination with the Antarctic quota, proved effective in dissuading most newcomers but not the irrepressible Dutch.[21] Many Norwegian whalers resented the crew law, because it limited their employment opportunities, and this resentment on the part of the Norwegian whalers allowed the Netherlands' state-subsidized company to get around the law simply by inviting Norwegian crews to join its expeditions in Cape Town, much to the irritation of the Norwegian government.[22]

But Norway and the United Kingdom's attempts at limiting the participants in Antarctic whaling were successful for the most part, considering that prior to the Second World War as many as ten countries engaged in some pelagic (as opposed to coastal) whaling while in the postwar period this number was mostly limited to five: the United Kingdom, Norway, the Netherlands, Japan, and the Soviet Union.[23] Given the postwar period's dearth of fats and skyrocketing oil prices, the idea of going whaling was a popular one in many countries and enthusiasm for whaling was itself at an all-time high. The hurriedly made plans of most governments for Antarctic hunting, however, were never realized:

> In 1945–50 it looked as if the whole world wanted to go whaling— Americans, Argentineans, Australians, Brazilians, Canadians, Chileans, Danes, Dutch, Finns, Germans, Italians, Japanese, Russians, Swedes, all had whaling plans, and practically everyone was thinking in terms of pelagic catching in the Antarctic. There was also talk of operating from shore stations in a number of places. On this occasion expansion ran into three obstacles that had not been encountered before: the Norwegian crew law, the I.W.C.'s 16,000 units, and the fact that the losers of the Second World War depended on the good grace of the victors.[24]

By trying to create a monopoly over pelagic whaling, Norway and the United Kingdom probably prevented things from becoming as bad as they could have been. But they were unable to limit the number of Antarctic whaling fleets to a number where the whaling olympic mode of hunting and the crippling amount of capital that it absorbed could be avoided. One result of the postwar return to Antarctic whaling and the whale oil boom was the beginning of the end of Norwegian and British involvement in pelagic hunting, since their privately run industries—unlike the state-sponsored Dutch and Soviet whalers—found it increasingly difficult to remain profitable in the face of increasing competition and dwindling stocks.[25] Another and more significant legacy of the period, however, was the creation of some major criteria I and II obstacles to regulation based on the majority of scientific opinion: a situation which would soon bring the Antarctic stocks of several species and the whaling industry itself to the brink of total collapse.

THE IWC AS AN INTERNATIONAL BODY

As noted earlier, the IWC was established at the 1946 Washington Conference as the administering body for the International Convention for the Regulation of Whaling. The commission's brief was to apply the rules and objectives set out in the convention to the whaling activities of its member states. By 1949, the year of the IWC's first meeting, held in London, a total of twelve governments had ratified the convention and become full members of the commission.[26] The main objectives of the convention, as set out in its preamble, are "to provide for the proper conservation of whale stocks and thus make possible the orderly development of the whaling industry."[27] These goals were determined in light of the convention's general recognition of the whaling industry's past excesses and also the global need for fats in the postwar era:

- Considering that the history of whaling has seen over-fishing of one area after another and of one species of whale after another to such a degree that it is essential to protect all species of whale from further over-fishing;
- Recognizing that the whale stocks are susceptible of natural increases if whaling is properly regulated and that increases in the size of whale stocks will permit increases in the number of whales which may be captured without endangering these natural resources;
- Recognizing that it is in the common interest to achieve the optimum level of whale stocks as rapidly as possible without causing widespread nutritional distress;
- Recognizing that in the course of achieving these objectives, whaling operations should be confined to those species best able to sustain exploitation in order to give an interval for recovery to certain species of whales now depleted in numbers. . . .[28]

The ICRW, upon which the IWC's existence and legitimacy as an international regulatory body is based, is made up of two distinct parts: the convention itself, which sets out the objectives, basic rules, and general codes of conduct for its members, and the schedule, which provides detailed information on exactly how whaling operations should be conducted in relation to the broader framework provided by the convention. Unlike the convention, the schedule is flexible, since it must allow for changes in commission policy in relation to the setting of quotas, the opening and closing of sanctuaries and hunting seasons, and the protection of species and stocks deemed endangered. However, the schedule cannot be altered without the consent of a three-quarters majority of the members present and voting at a meeting of the commission.[29]

Thus, the main purpose of the IWC's annual meetings is to review the existing schedule and to make changes where necessary in accordance with the wishes of the three-quarters majority of voting members (each contracting government is represented by a commissioner who is entitled to one vote), so long as these changes are deemed to reflect the objectives of the convention and follow the provisions that its various articles set out. A frequently criticized characteristic of the ICRW, however, is the often-vague nature of its provisions and also the "escape clause" that it provides for members not wanting to adhere to schedule amendments that have achieved the required majority vote. Perhaps the most controversial of the ICRW's articles is Article V, which states in paragraph two,

> These amendments of the schedule (a) shall be such as are necessary to carry out the objectives and purpose of this Convention and to provide for the conservation, development and optimum utilisation of the whale resources; (b) shall be based on scientific findings; . . .[30]

Paragraph three of Article V goes on to say:

> Each of such amendments shall become effective with respect to the Contracting Governments ninety days following notification of the amendment by the Commission to each of the Contracting Governments except that (a) if any government presents to the Commission objection to any amendment prior to the expiration of this ninety-day period, the amendment shall not become effective with respect to any of the Governments for an additional ninety days; (b) thereupon, any other Contracting Government may present objection to the amendment at any time prior to the expiration of the additional ninety-day period, . . . and (c) thereafter, the amendment shall become effective with respect to all Contracting Governments which

have not presented objection but shall not become effective with respect to any government which has so objected until such date as the objection is withdrawn.[31]

The contradictions and ambiguities evident in the convention requirements cited here—the convention's conservation and optimum utilization objectives and the requirement they be pursued on the basis of scientific findings *in addition to* the provision for governments not to be necessarily bound by commission policy—provided the basis for much of the IWC's inability to conserve whale stocks during the commission's first two decades. These aspects of the ICRW also are relevant to many of the problems the commission continues to experience today, particularly in terms of how the IWC membership has interpreted scientific advice and chosen to deal with it. The ICRW's simplistic requirement for schedule amendments to be "based on scientific findings" ignores both the inevitability of differing scientific findings being presented and the even more fundamental issue of what kind of scientific advice should be given priority when conflicts arise (i.e., how should the commission choose between science attempting to demonstrate that hunting is justified and science claiming that protection is required). No unequivocal statement as to how these issues should be dealt with is provided by the convention since its objectives also are less than clear thanks to its use of phrases such as "optimal utilization" and "the orderly development of the whaling industry" that can be understood to mean a large number of different things.

The problems of interpretation caused by the ICRW's vagaries have been further compounded by the IWC being unable to enforce its policies in the absence of unanimous agreement, since the ninety-day objection provision allows members to ignore schedule amendments when they believe such changes are not in their interests. Furthermore, the unclear wording of the schedule allows dissenting members to justify such noncompliance by arguing alternative interpretations of the convention. Thus, as the arguments and examples of this study later will show, it has been possible since the IWC's inception for various governments to pursue entirely different goals in relation to other IWC members while at the same time claiming to be acting in accordance with the ICRW.

Many have criticized the ICRW's vague wording and, in particular, the ninety-day objection provision, which severely limited the IWC's ability to enforce its decisions.[32] Moreover, with the advantage of hindsight, it is clear that the IWC's failure to adequately conserve stocks during the postwar period was, at the very least, exacerbated by these problems. But according to Steinar Andresen, "The goals of the ICRW—although both vague and maybe inconsistent—were an attempt to compromise between the different interests represented in the IWC. Generally it was not the goals of the ICRW which were wrong, but the way they were implemented."[33] And as others have noted, vaguely worded conventions and treaties are not uncommon and are often essential to the formation of international regimes, as are the inclusion of so-called escape clauses.[34] Former IWC general secretary and SC member Ray Gambell, for example, goes one step further: "This escape clause [the ninety-day objection rule] was designed to allow governments not to be bound by regulations which they consider to be detrimental to their own best national interests . . . but it is doubtful if the Convention itself could have been approved without such an arrangement."[35]

In other words, while the convention's lack of clarity and inclusion of an escape clause has contributed to the IWC's lack of effectiveness in terms of conservation (and continues to do so today), it is highly unlikely that a more binding version would have been ratified by the governments that have had the most impact on the whaling industry's development. Therefore, the real problem confronting the convention's implementation of conservation initiatives has not been its wording and contents so much as the unwillingness of its members to compromise and refrain from interpreting the ICRW's provisions and intentions only in ways that further their respective ambitions.

> This [the depletion of whale stocks] is not the fault of the Commission as such, but is because a number of members have refused to recognise facts and have argued that they should be allowed to catch as long as there is anything to be caught.[36]

This statement by historians Johan N. Tønnessen and Arne Odd Johnsen accurately summarizes the single-mindedness that characterized the attitudes of many IWC members toward quota reductions in the postwar period. But it also ignores the important link between the refusal of members "to recognize facts" and their ability to argue in favor of continued catching. The uncertainty that has been (and remains) so prevalent in cetacean science means that "the facts" never have been beyond dispute so long as someone has had something to gain from disputing them, even when such facts could be empirically supported by sudden drops in the number of whales caught in spite of the greatly increased capabilities of the whaling fleets or increasing numbers of younger whales being taken.

The important question, then, is why opponents of lower quotas were able to argue so successfully against "the facts" presented by scientists in the IWC in support of reduced catches. The answer lies with the unavoidable uncertainty that accompanies scientific advice, due to the problematic nature of identifying and interpreting "the facts" in the first instance (i.e., in the process of formulating a given theory), and then, in the second instance, the various political motivations that ultimately determine the acceptance or rejection, and therefore the veracity, of scientific evidence within the policy-making process.

Put simply, "the facts," as presented by a body of research, can be whatever people and policy makers want them to be (i.e., reliable or highly questionable) in relation to the course of action they believe to be the most advantageous (or that provides the most "utility") This is especially so in a regime like the IWC, where priorities are unclear and the influence of one set of interests usually is heavily favored by the majority (i.e., the proindustry bias of the IWC membership during the 1950s). Science, therefore, played an important role in the IWC membership's various policy machinations not simply because of its perceived ability to describe reality but because of the uncertainty inherent to science, particularly marine science, which frequently allows various forms of reality to be argued for.

SCIENTIFIC ADVICE IN THE IWC

Although it is now one of the IWC's three permanent committees—the Scientific Committee (SC), the Technical Committee (TC), and the Finance and Administrative Committee—the SC was originally formed as part of a Scientific and Technical Committee and relied on scientists working in ad hoc subcommittees for recommendations it would then discuss and present to the commission.[37] The SC and TC did not become separate committees until 1951.[38] This allowed the SC to focus on the reports it received from its subcommittees rather than being distracted by other issues such as the whaling regulations of member countries, catching methods, and infractions, which became solely the brief of the TC.[39]

The SC's structure has remained basically unchanged since the creation of a standing Scientific Sub-Committee in 1955 (of which there are now several), although its role was "considerably expanded" by the adoption of new procedure rules for the SC in 1981.[40] Notwithstanding the various changes to meeting schedules and rules of procedure that have occurred over the years, the functioning of the SC has remained basically the same.[41] The plenary sessions of the IWC at each annual meeting have directed the SC to pursue particular research issues and problems, which it then delegates to its various subcommittees approved by the commission.

The reports from these subcommittees are then given to the SC for deliberation and inclusion in part in the SC report to the commission, which is included in the annual report for the following year. The SC's recommendations are first considered by the TC, which acts, in addition to its other duties, as a form of preliminary plenary session since most member governments maintain representatives at the TC meetings. The SC recommendations are finally discussed and voted upon at the plenary session for that year's annual meeting.

Scientific research in the IWC, in spite of the important role given to it by the ICRW, mostly remained on the fringes of the commission's policy making until the late 1950s, when the threat of whaling's imminent collapse was sufficiently convincing—as a result of increasingly poor catches—for enough of the pelagic whaling nations to prompt at least a partial change in the commission's perception of uncertainty and the role of scientific advice. The relatively minor role afforded to scientific advice in practice was mostly due to the IWC's strong proindustry bias and a weak cetacean science that could not counter industry dominance. But the ineffectiveness of scientific advice during this period can also be linked to how the scientists were organized within the commission and the limits imposed upon them by the commission.

The IWC scientific community was small and somewhat isolated during the 1950s, and while many of its members were competent in cetacean biology, few had any expertise in the important fields of statistical analysis and population dynamics. As SC member Tore Schweder has noted, the postwar period saw significant progress in statistical methodologies and fisheries science, culminating in Raymond Beverton and Sidney Holt's landmark work *On the Dynamics of Exploited Fish Populations* in 1957, and these fields have since provided the basis for the development of cetacean management science.[42] The IWC's Scientific Committee, however, was seriously restricted in its ability to take advantage of such expert advice by budgetary pressures that, in addition to thwarting attempts to include members of the broader scientific community in its deliberations, forced its own members to meet "somewhat irregularly" during the 1950s and also delayed publication of the SC's reports to the commission until 1955.[43]

Since becoming a separate committee, the SC's deliberations ostensibly have been focused on only the biological issues pertaining to management as described by Article IV of the convention,[44] but several instances during the 1950s demonstrated that the SC's advice on biological issues was often tempered by nonbiological factors. At the IWC's sixth meeting in 1954, a proposal for quotas based on species rather than the BWU in the SC subcommittee was supported on the grounds "it would be a great advantage." But the committee agreed not to put the proposal forward after a Norwegian member pointed out that "there would be great practical difficulties in operating such separate quotas."[45]

In 1955, alarmed by the rapid decline in blue and fin stocks, the SC believed that, in addition to separate quotas, a drastic cut in the quota from 15,500 to 11,000 BWU was needed. But as in the previous year, the scientists realized the commission would not accept such a sudden reduction due to the hardship it would cause the whaling companies. Instead, they recommended reducing the quota incrementally, starting with 14,500; but even this watered-down proposal was rejected by the commission.[46] These developments eventually resulted in the SC chairman, N. A. Mackintosh, being strongly criticized by the New Zealand commissioner at the following year's meeting:

> From one meeting to another, [the New Zealand commissioner] asserted, one gained a stronger and stronger impression that the [Scientific] Committee took into consideration factors that were anything but purely scientific. . . . He put a direct question to Mackintosh: What was the *scientific* reason for proposing such a small reduction now and a larger one later? Mackintosh was openly forced to admit that the Committee had allowed itself to be influenced by factors that were irrelevant to it. On the other hand, the Committee had felt that it should take into consideration what was feasible, and not what it considered *should* be done on the basis of a biological view of what might be desirable.[47]

During the 1950s, the SC clearly was working under difficult conditions. In addition to scant data on stock numbers, which relied almost entirely on information from catches, its members were handicapped by their own lack of expertise in what was largely an undeveloped field (cetacean management). The SC also was prevented from bringing outside experts into its meetings due to insufficient funding. The question of why SC members allowed their advice to be so heavily influenced by what they thought the commissioners would or would not accept is an important one—particularly in terms of the implications it has for views, like the New Zealand commissioner's, that science is or should be politically neutral—and can largely be answered by looking at the SC's weak position and the dominance of the commission's criterion I priority (i.e., management for maximum financial return in the shortest time possible). Indeed, some observers believe the SC should have been more forceful and lay much of the blame for the excesses of the 1950s at the feet of the commission scientists. David H. Cushing, for example, writes, "the ultimate blame [for excessive catches] must lie with the scientists, who ignored what had happened in the past. Perhaps they were overwhelmed by their lack of exact information and lacked the will to give good advice without it."[48]

Cushing's assessment, however, is a little harsh given the constraints the SC was facing. The historical record also shows that the majority of scientists did indeed try to warn the commission of the extent to which stocks were being depleted. Had the commissioners been prepared to listen to the majority of scientific opinion concerning the status of the blue, fin, and humpback stocks, it seems likely a more conservative approach to the setting of quotas would have been taken.

The majority of member governments clearly were not prepared to take on any advice that would endanger profits and for this reason a more forceful stand by the scientists was unlikely to have made much difference. On several occasions, the SC members, with the exception of the Dutch scientists, made their belief in the need for much smaller quotas quite clear. But the growing competition and financial pressures among the Antarctic nations meant there was little chance their advice would be heeded. The Antarctic nations in effect had put themselves in a situation where they simply could not afford the kind of scientific advice that was being offered. And so they rejected it.

The advice of the SC, for this reason, was undermined by the specter of uncertainty (which is always present regardless of the available data and methods), and the extent to which governments made an issue of it depended entirely on how compatible the committee's advice was with IWC governments' existing priorities and goals. In the IWC during the 1950s, scientific advice advocating caution and conservation at the risk of profits obviously was unwelcome. For this reason, rather than simply because of data and methodology shortcomings, the assertion of uncertainty became an important tool in delaying the adoption of lower quotas. This situation is well illustrated by the conduct of the Dutch scientists in the SC during its deliberations on the status of the Antarctic's fin whale stocks.

SEEING ISN'T ALWAYS BELIEVING: SCIENTIFIC UNCERTAINTY IN THE IWC 1949–59

The IWC's founding charter, the ICRW, with its emphasis on conservation, optimal utilization, and scientific findings, indicates that the postwar era of commercial whaling indeed held promise for a more restrained and responsible approach to cetacean management. But as discussed earlier, the postwar whale oil boom, caused by the fats shortage in Europe and other regions, encouraged a rapid return to pelagic and shore-based whaling under circumstances that quickly placed immense pressure on stocks in the Antarctic and elsewhere—leading to the accelerated depletion of stocks and the demise of commercial whaling. The pelagic whaling fleets, in effect, were hunting themselves out of business.

The major obstacle to the IWC realizing its goals undoubtedly was the process of overcapitalization set in motion by the postwar shortage of edible fats and oils. More conservative management initiatives were clearly in conflict with the dominant criterion I imperative for whaling during the 1940s and 1950s (i.e., the high utility of catching as many whales as quickly as possible), which was legitimized by the widespread demand for fats. Even if the IWC had accepted that there was greater long-term utility in reducing catches and conserving stocks, the adoption of a more cautious approach to hunting still would have been rendered problematic by political pressure stemming from the already established global need for increasing quantities of edible oils and fats.

It can of course be argued that there is little utility in rapidly exhausting a resource that was of such critical importance.[49] But the dominance of the industry and its criterion I imperatives, in addition to the generally weak position of cetacean science and in particular the IWC Scientific Committee at the time, left conservation advocates within the IWC almost entirely without influence during the 1950s. Indeed, the shortcomings of cetacean science effectively made it easier than it otherwise would have been for members of the commission to downplay the risk of overexploitation by invoking scientific uncertainty.

It is difficult to imagine that even the strongest opponents of lower quotas did not privately suspect the stocks were being excessively taxed, given the signs. The majority of IWC scientists held this opinion, the blue whale catch sharply dropped in the early 1950s, and increasing catching efforts were required by the late 1950s—in spite of far more efficient equipment—to obtain a significantly smaller number of BWU than before.[50]

Problems with the existing data and the inability of scientists to confidently answer critical questions concerning stock numbers, reproductive capabilities, and ability to support commercial hunting were, however, nothing new. The extent to which cetacean science and management were hamstrung by a dearth of data and accepted population analysis techniques is well illustrated by the ongoing use of the BWU and also by the arbitrary way in which the postwar limit of 16,000 was first decided in 1944 and then reconfirmed in Washington two years later.

As Tønnessen and Johnsen have observed, the IWC's three leading scientific advisors of the period—Remington Kellogg of the United States, N. A. Mackintosh from the United Kingdom, and Norwegian scientist Birger Bergersen (of whom only two were marine biologists; Kellogg was a paleontologist)—were essentially guessing what the limit should be and seemed more concerned with the imposition of a limit of some kind than they were with the actual number it might involve:

> How did they arrive at the exact figure of 16,000? At the 1944 conference it was stated that even though this amount could not be caught during the first season, the intention was "to prevent the present situation being exploited for unchecked building of new floating factories," and secondly, "because there is a desire to create a precedent for total limitation in the future." . . . The Norwegian delegate gave an account of these plans at the conference: "We proposed therefore that for this season a total limitation of catching should be established. A total catch of 16,000 BWU was agreed on, i.e. 1.6 to 1.7 million barrels, which it was calculated could be extracted from about 20,000 whales, about half the yield immediately before the war." He relates that he proposed 16,000, instead of 15,000 or 20,000 as proposed by Kellogg and Mackintosh, as it "seemed to be rather more reassuring." In this rather fortuitous fashion was this fatal figure arrived at! This impression of chance was further confirmed by the fact that the three gentlemen in question [Kellogg, Mackintosh, and Bergersen] agreed that "the figure for the total quota was of minor importance," the principle of total limitation being the most important.[51]

Thus, from the outset, the IWC's method of basing its policies on scientific advice already had been skewed by a preponderance of uncertainty over stock numbers, and also the industry's need to produce a yield of whale oil capable of meeting international demand. Obtaining a better idea of stock numbers was of secondary importance at the time, since the priority was to provide edible oils while establishing a limit on hunting, regardless of how arbitrarily that limit may have been arrived at.

Under these conditions, the 16,000 BWU limit remained unchallenged for the most part until 1951 when, after the quota had been reviewed in the Scientific and Technical Committee's agenda at the IWC inaugural 1949 meeting,[52] the SC concluded that the quota was probably too high but failed to recommend any adjustment.[53] As mentioned earlier, the SC and its subcommittees were ineffective in convincing the IWC commissioners that large reductions in the quota were needed, largely because of commissioners' unwillingness to listen and also because some scientists in the SC disputed the threat of overhunting. Sir Gerald Elliot, chairman of Britain's largest whaling company at the time, Christian Salvesen PLC, recalls the circumstances facing science-based conservation initiatives in the IWC during the postwar period as being very unfavorable:

> There had been plenty of work done by British and Norwegian scientists on Antarctic stocks in the 1930s but their conclusions had been very tentative. All they knew, as everyone could see from the catch figures, was that the blue whales, preferred for their size, were reducing in the catch relative to fin whales, indicating a scarcity, and that greater catching effort was not taking more whales. Professor Mackintosh had initiated some primitive counts of whale sightings . . . but evidence from these was hardly enough to satisfy governments and whalers. Until the middle of the 1950s there was not even an accepted method of telling the age of whales, an essential element in population dynamics. So the pleas of the Scientific Committee . . . that the initial 16,000 unit quota was too high and should be cut received a rough reception in the plenary sessions of the IWC.[54]

From 1949 to 1952, the major issues in the IWC included calls for more research on the Antarctic stocks and the lifting of the Antarctic ban on humpbacks,[55] the prevalence of catches significantly exceeding both the BWU quota and the species quota that had been set for humpbacks,[56] and the beginning of a series of warnings made by IWC scientists that the 16,000 BWU quota was too high.[57] Continuing concern over the status of stocks led to a proposal to introduce individual stock quotas, first made at the 1951 meeting by Mackintosh, then SC chairman, and repeated in London in 1952; both attempts were unsuccessful.[58] By the 1952–53 season, catches of blue and humpbacks had dropped significantly,[59] prompting speculation by scientists that blue and fin stocks were in fact being depleted.

The 1953 meeting, again held in London, was notable for two important reasons: (a) the SC for the first time recommended lowering the quota (to 15,000 BWU); and (b) Dutch scientist E. J. Slijper, who would become a consistent opponent of the SC's findings and recommendations over the next seven years, began questioning the committee's majority opinion that the quota was too high. In addition to advising a reduction in the quota at the 1953 meeting, the SC also recommended the complete protection of the blue whale and restrictions on the operations of factory ships. The Technical Committee opposed the latter two proposals but endorsed lowering the quota to 15,500 BWU. The Netherlands, however, rejected a reduction to 15,000 on the grounds of uncertainty based on arguments Slijper had put forward in the SC.

The main source of data on whale numbers and the status of various stocks during the 1950s was the composition of catches (sex, size, and age) and also the catch per unit of effort (CPUE) of the catcher boats, or catch per catcher day's work. The CPUE was used extensively as an indication of population abundance until the late 1970s, when it was noticed that decreases in the CPUE were only likely to become apparent after stocks had already been heavily overexploited. The CPUE's use as a gauge for population number estimates, according to Allen, is based on the following two assumptions: (a) catch for given effort under given conditions is proportional to population size; and (b) the catch from a given population is proportional to the effort index.

> Until recently [i.e., the late 1970s] whale population estimates have been based generally on the assumption that the catch per catcher-day's work—that is, the average number of whales caught by a catcher in a day—is, with a correction for vessel efficiency, proportional to the size of the population.[60]

Norwegian scientist and IWC chairman Birger Bergersen's warning at the 1953 SC subcommittee meeting that the "stocks of blue and fin whales considered as a whole show unmistakable signs of depletion" was based on analysis of catch composition and CPUE in particular areas of the

Antarctic. Slijper's argument against Bergerson's conclusion, which otherwise had the full support of the SC,[61] was based on what he believed to be the unreliability of a catcher day's work as an accurate unit of effort due to the increasing competition between catcher boats in the prevailing whaling olympic hunting environment.[62]

Thus, according to the Dutch, decreasing CPUE indications were just as likely the result of reduced efficiency caused by increasing competition, rather than from an overall drop in whale numbers—more ships hunting the same stocks could mean wasted time and effort through two or more ships chasing the same whale but only one actually catching it. The Dutch delegation, in a memorandum to the SC, argued against warnings from the Special Scientific Sub-Committee[63] that the current catches were probably too high, citing uncertainty over the reliability of both CPUE indicators and the evidence of smaller animals in the catches. The Dutch scientists also raised the possibility that more blue whales would be found in the ice, rather than outside of it, since "it is common knowledge that the majority of blue whales live in the ice."[64]

According to Schweder, Slijper also used population estimates taken from Mackintosh and Norway's J. Ruud to make his own calculations on the fin whale replacement yield, which conveniently came out to be the same figure as the existing annual take of 24,000 fin whales.[65] The IWC, however, voted in favor of a quota reduction—though only by 500 BWU rather than the 1,000 recommended by the SC—while the Dutch government opposed the reduction on the grounds of uncertainty.[66] In its memorandum, the Dutch delegation concluded that "at the moment there is no sufficient scientific evidence for the conclusion that it is absolutely necessary to reduce the number of 16,000 Blue Whale Units."[67]

Concern for the state of the Antarctic populations, particularly the blue and fin stocks, continued at the 1954 meeting in Tokyo, where the commission acknowledged the possibility that some species were being overhunted. The problem anticipated earlier by scientists such as Mackintosh, Bergersen, and Ruud was that the apparent decreasing catch of blue whales would lead to extra pressure being placed upon the fin stocks, following the process of sequential depletion that had so far characterized commercial whaling. These concerns were justified since, as the SC subcommittee noted at its 1955 meeting, the catch of fin whales had increased from 17,474 in 1950–51 to almost 26,000 by the 1954–55 season.[68] It was also for this reason that the ban on humpbacks had been lifted in 1949 (to reduce pressure on the blue and fin whale populations) and was not reinstated after two seasons as had been originally agreed.[69] In the Chairman's Report of the Tokyo meeting, the commission explicitly recognized the threat of overhunting and appeared to make the adoption of a lower quota contingent only upon the existence of sufficient evidence that stocks were in decline:

> Having regard to the scientific advice at their disposal and to the catch statistics covering whaling operations in the Antarctic, the Commission expressed the opinion that it may soon become necessary to restrict more severely the Antarctic catch of blue whales, while guarding at the same time against a corresponding increase in the catch of fin whales. This alone would involve a reduction in the total permitted catch in the Antarctic. If there should be clear signs of depletion of fin whale stocks also, the Commission believe that a further and very substantial reduction of the total permitted catch should be made at once.[70]

Indeed, by 1955, the majority of IWC scientists already had agreed that blue whale populations in the Antarctic, North Pacific, and North Atlantic and also fin whales in the Antarctic were under extreme pressure from hunting. With regard to the Antarctic blue whale stocks, the subcommittee scientists warned that they took "a grave view of the condition of this species. It would appear that the stock is now only a fraction of the original population, and its powers of recovery might already be found to be largely lost even if it received total protection."[71]

Similar fears were voiced over the condition of the fin whale, based on mortality rate calculations (taken from age determination using baleen plates) made by Ruud and supported by British scientist R. M. Laws's mortality calculations using ovaries analysis: "The conclusion is that the total mortality rates, including mortality from whaling, over a series of years, have been in excess of the maximum rate which would permit the maintenance of a stable population, and therefore that the stock of fin whales is in the process of depletion."[72] The subcommittee, therefore, "strongly recommended" at its 1955 meeting that a proposal for a quota reduction to 14,500 units be put on the agenda for the 1955 plenary meeting in Moscow.[73]

Slijper reportedly "was not disposed to dissent out of hand from this recommendation." He was not, however, "prepared to endorse it wholeheartedly" either. Slijper refused to accept Ruud's report until he could study it further and also argued that if catching were concentrated on older whales by raising the size limit, there would "on balance be a net gain in recruitment of young whales to the stock." However, the subcommittee was skeptical of Slijper's assertion, since it depended upon assumed mortality rates for different ages and in particular upon the maximum ages of whales—issues for which little or no information existed. Another problem was that Slijper's opinion assumed an undisturbed stock, which clearly did not exist. The general feeling in the subcommittee was that Slijper's suggestion would not improve the current situation, since "any beneficial effect of raising the size limit again would be long delayed."[74]

With the threat of decreasing quotas looming at the Moscow meeting, and the likely appearance of further evidence in support of Laws and Ruud's findings (as was later presented by Norwegian scientist P. Ottestad at the SC subcommittee's meeting in March 1956),[75] the Dutch scientists intensified their opposition to claims that the fin whale was being overhunted. They suggested at the SC meeting that fin whale numbers probably were increasing rather than decreasing. Slijper and fellow Dutch scientist E. F. Drion criticized Laws and Ruud's findings on the grounds that the samplings used may not have been random. In addition, Slijper and Drion cited their own calculations as equal reason to believe the Antarctic fin whale population was increasing. According to Schweder, "That the fin whale stock might be on the increase despite the catch of some 24,000 a year, must have caused some lifted eyebrows. . . . [Slijper and Drion's] calculations were based on assumptions that contradicted the fact that fin whales have a finite carrying capacity."[76]

The 1955 meeting in Moscow was largely taken up with the issue of reducing the 15,500 BWU quota to 14,500 units as recommended by the SC subcommittee. The eventual proposal voted on by the commission was divided into two parts: an initial reduction of 500 units to 15,000 BWU for the 1955–56 season; and a further 500-unit reduction to 14,500 thereafter.[77] Both amendments to the schedule achieved the required majority (eleven of the fifteen commissioners present voted in favor of both), with only Panama and the Netherlands voting against both parts and Japan voting against the first while the United Kingdom opposed only the second part.

This apparent victory for conservation was weakened by the opening of the Antarctic Sanctuary as an apparent concession for the reduced quota and was, in any case, short lived.[78] The initial reduction of 500 BWU received no objections and was then adopted into the schedule, thereby becoming binding for all member governments. But the second, and longer-lasting, reduction received less support. The Dutch commissioner lodged an objection against the quota of 14,500 for the following two seasons (1956–58), which effectively meant no quota would apply to the Netherlands. Fearing the disadvantage this would create for their own fleets, other governments quickly followed the Dutch lead:

> Eventually the first reduction, i.e. 15,500 to 15,000 for the season 1955–56 came into operation as from 8 November 1955, but when the second reduction was referred to Contracting Governments, in accordance with the required procedure, the Netherlands Government objected. Their objection was subsequently followed by the following Governments, namely the UK, Panama, South Africa, Norway, Japan, USA and Canada. The result was that the second reduction did not come into force until 7 March 1956 and was not then enforceable against the objecting governments mentioned. It should be mentioned that the bulk of scientific opinion in the Commission was in favour of still greater reduction.[79]

Almost all the governments of the major pelagic fleets except the Soviet Union felt compelled to copy the Dutch stance. Thus, in effect, the Dutch government stymied the plan to incrementally reduce the quota by 500 units until a sustainable level was reached (a level that many scientists still believed to be around 11,000 units).

The issue of whether lower or higher quotas were justified depended primarily on the fin whale stocks, since they were now the mainstay of Antarctic whaling—a point already noted by the SC subcommittee prior to the IWC Moscow meeting when its members concluded "the whole Antarctic whaling industry is virtually dependent on healthy stock [sic] of fin whales."[80] Thus, the subcommittee's further observation, based on unanimous agreement, "that definite signs of depletion, such as indicated by [Laws and Ruud's research], point to a very dangerous situation," meant that any opposition to lower quotas would need to be based on arguments capable of demonstrating at least that fin stocks were not necessarily being depleted.[81]

Equivocation and the emphasizing of uncertainty provided readily available means of casting the required level of doubt over the SC's conclusions, especially given the very limited resources the committee had to work with, the subsequent weakness of its evidence, and also the general reluctance within the IWC to risk profitability through smaller quotas unless absolutely necessary.

The attitude within the commission that the industry's interests should not be compromised without more convincing arguments from the SC was made clear by the British commissioner at the Moscow meeting: "the position of the whaling industry should be fully taken into account and balanced with scientific requests as far as possible . . . unless scientists are able to convince us that a most serious situation will develop."[82]

The main perpetrators of uncertainty (aimed at undermining calls for reduced quotas) were undoubtedly the Dutch commissioner (and IWC chairman in 1957–58), G. J. Lienesch, at the plenary session level and Slijper and his assistant Drion in the scientific committees. And their reasons for doing so were probably part of the Dutch government's plans to intensify its whaling interests in the Antarctic, as indicated by the introduction of a new Dutch floating factory in 1955.[83] After the 1955 meeting, the intentions of the Dutch delegation were clear. Their conduct in Moscow and at subsequent meetings made the Dutch scientists almost synonymous with opposition to reduced quotas and with the invocation of scientific uncertainty:

> The Netherlands consistently refused to accept the necessity of the proposed reduction every season. In opposition to the unanimous views of all the other ten members of the Scientific Committee, that a catch of 25,000 fin whales a year would rapidly result in the extermination of stocks, the Netherlands asserted that there were equally good grounds for maintaining that stocks of fin whales were twice as large as biologists calculated, and were therefore capable of sustaining a yield desired by the Netherlands [i.e., at least 16,000 BWU].[84]

At the 1956 meeting in London, the Soviet Union delegate announced that his government also would not adhere to the 14,500-unit quota, due to the objections made by other members to quota reduction. The Soviet announcement followed the premeeting distribution of an IWC circular to the commissioners that made clear the broader implications of the Dutch objection to the quota scheme in Moscow a year earlier:

> It is perhaps desirable to explain that the 15500 blue whale unit limit has been reduced to 15000 in respect of season 1955/56 and to 14500 units thereafter, but seven countries objected to the further reduction to 14500

units and are not therefore bound by this figure, which however is binding on the 10 non-objecting countries. It happens however that the seven objecting countries, as things stand, are not in fact bound by any limit at all after 1955/56, and it is essential that this anomalous position should be put right without delay."[85]

In the same document, written by the secretary to the commission, A. T. A. Dobson, it was suggested that the problem could be solved "by omitting any reference to the 14500 units and leaving the 15000 limit to operate for the future."[86] This proposal from within the commission itself gives a clear indication of just how much concern the situation had generated and also the extent to which priority was given to the interests of the industry.

The commissioners, however, voted to amend the schedule so that the 14,500 quota would only remain in force until the end of the 1956–57 season before reverting to 15,000 BWU for the following season.[87] The Dutch commissioner again dissented, in opposition to the lower catch, claiming "there was not sufficient evidence to show that, on the basis of the present calculations, the proposed reduction is necessary."[88] On this occasion, however, no objection was lodged.

At the 1957 meeting, again in London, the repeated warnings of the SC subcommittee that the Antarctic fin whales were in decline appeared to finally be taken seriously by the commission, with the notable exception of the Netherlands government, which continued to claim there was not sufficient evidence to justify the subcommittee's conclusions. This time the commissioners from the United Kingdom, the Soviet Union, Norway, France, and Japan all agreed that "although there was no conclusive proof of a heavy decline in the stock of fin whales, the balance of evidence justified a warning that the number of whales taken annually in the Antarctic was dangerously high."[89] The result of this was for the commission to vote in favor of returning to the earlier plan of keeping the quota at 14,500 units for the 1957–58 season. And while support for this amendment was not unanimous, no formal objections were made at the meeting.[90]

The 1958 meeting in The Hague followed the close of the 1957–58 season, which meant that the quota had automatically reverted to 15,000 BWU prior to the meeting. Again, the majority of commissioners heeded the advice of the SC, again with the exception of the Netherlands representative. The majority agreed to keep the 14,500 limit for another year (i.e., the 1958–59 season) and to amend the schedule accordingly. On this occasion, however, the Netherlands objected to the amendment and, as at the 1955 meeting, the Dutch objection was then followed by objections from "the other four governments with whaling fleets in the Antarctic," namely the United Kingdom, Norway, the Soviet Union, and Japan. Subsequently, the "effective quota" for the 1958–59 season remained at 15,000.[91]

A common theme of the IWC's meetings from 1953 onwards was the constant opposition from the Dutch delegation to any reduction in catches, based on Slijper and Drion's criticisms concerning the veracity of the evidence used by other scientists to argue in favor of smaller catches. Slijper and Drion's modus operandi was also remarkably consistent: they claimed that the SC and its subcommittee's conclusions were based on "insufficient evidence" and that various interpretations were possible—thereby implying that warnings of excessive hunting were largely speculative.

At least some of Slijper's contemporaries, however, believed that his constant criticism and questioning of the work of other committee members betrayed both a lack of understanding of the issues and, more importantly, a desire to justify the existing quota regardless of any evidence to the contrary. Schweder writes,

> According to Laws, the prevailing opinion within the Scientific Committee was that Slijper lacked competence in the field of population dynamics and statistics. Slijper kept posing new questions, apparently not to enhance understanding, but rather to foment uncertainty. It is extremely time consuming to address such questions, and the Scientific Committee was often unable to resolve such matters in the course of one meeting. The questions, therefore, were left unanswered, and additional questions were posed at the next meeting. One of Slijper's hypotheses was that the fin whale stock could be increasing in the face of heavy harvesting. Laws recalls that when it was pointed out to Slijper that the hypothesis was inconsistent with the existence of a self-regulatory mechanism, it became obvious that he did not understand the concept of density dependence in population dynamics.[92]

After 1958, however, concerns over the size of the quota and the reliability of the SC's warnings were overshadowed. Shortly after the meeting in The Hague, the Netherlands, Japan, and Norway posted notices of withdrawal from the IWC following the failure of the five Antarctic whaling governments to agree on an allocation of national quotas to replace the one quota system. Norwegian and British companies had adhered to national quotas prior to the war but such quotas were not advocated in the ICRW. This was largely because of the prevailing free-trade ethic and freedom of the seas doctrine in the postwar years, particularly favored by the United States, and also the Norwegian and British whalers' belief that they would be able to maintain their prewar dominance of whaling.

The possibility of investment costs spiraling out of control due to the whaling olympic mentality had not occurred to the older whaling nations at the end of the war, since they assumed that the number of participants would remain limited. Norway and the United Kingdom also failed to anticipate the impact that entrants such as the Netherlands, the Soviet Union, and the Japanese would have on the industry's profitability. Thus, it was from a British proposal that the plan for national quotas was first mooted in the IWC at the 1958 meeting, after discussions between British and Norwegian companies had begun as early as 1955.[93]

The five Antarctic whaling members—Norway, the United Kingdom, Japan, the Soviet Union, and the Netherlands—met in London after the 1958 IWC meeting to discuss allocation of quotas. These discussions needed to occur outside of the IWC due to Article V of the ICRW prohibiting the setting of national quotas.[94] As a result, the IWC had no influence over these discussions in spite of the huge impact they would have on the commission and Antarctic whaling. The central issue behind the proposal for national quotas was the imbalance between the economic pressures of the whaling industries, caused by the large amount of infrastructure in use, and the biological limitations of the whales themselves, which was under dispute.

Given the more secure financial positions of the state-run Soviet and state-subsidized Dutch industries, in addition to the benefits of strong domestic demand for whale meat in Japan enjoyed by the Japanese companies, the British and Norwegian companies—which had neither state support nor strong domestic markets for whale meat—were clearly at a disadvantage and were, therefore, the worst affected by the uneconomical nature of the competitive single-quota system. The intention, then, was to reintroduce a balance to whaling by reducing the number of ships, floating factories, and other materials used by the whaling companies to a level that was economically defensible in terms of the number of whales that could be sustainably hunted each year.

Thus, even the allocation of national quotas remained reliant on the adoption of a total annual catch that was deemed suitable by the Antarctic whaling countries. It was agreed at the first meeting in 1958 that this figure should be the quota adopted under the ICRW. But because the Netherlands continued to insist a total of 16,500 BWU (including 500 humpback units) was possible, while the other participants believed 15,000 to be the absolute maximum, the question of whether or not fin stocks were in decline remained in dispute. The implications and outcomes of this situation, in addition to the more influential role given to scientific advice in the 1960s, will be examined in chapter 3. But it is clear that by the end of the 1950s, the situation facing all concerned in the IWC, and in particular the Antarctic whale populations, had clearly deteriorated into a major crisis.

Unfortunately, given the circumstances and environment the IWC was operating in during its first decade, it is hard to imagine how a different outcome could have been possible. Many critics of the IWC and its membership's actions during this period, including Tønnessen and Johnsen, Schweder, and Birnie, have suggested—either explicitly or implicitly—that the mismanagement of quotas and the collapse of stocks in the Antarctic and elsewhere could have been avoided if only the commissioners had listened to the advice of the scientists. Others, such as Cushing, believe the scientists should have taken a more aggressive stance and presented their findings more forcefully. Elliot, a British whaling industry representative at the IWC, takes a slightly different view, placing most of the blame upon, firstly, the Soviets and, secondly, the Dutch and also stating that "the Norwegians and British come out quite well in an historical assessment."[95]

On the surface at least, the historical record of the period, most notably the IWC documents and opinions from those who were there, generally support the view that the worst excesses of the period could have been avoided if the majority supported advice of the SC and its subcommittees had been acted upon. But in order for this to have been the case, an almost ideal international environment—one that probably has never existed—would have been required: an environment where governments do not elevate perceived national interests above all others and where tangible short-term benefits are forsaken for more obscure and less certain long-term benefits.

This ideal environment would also require scientific advice to be able to unequivocally and objectively predict outcomes to the satisfaction of all concerned. But, as most people would agree, such a scenario remains impossible, even fifty years after the fact. So it seems somewhat pointless to argue in hindsight that a given outcome should have occurred when the environment needed to produce that outcome is itself little more than the product of idealist thinking. This is not to say that there are no lessons to be learned from past experience. The question is whether we are learning the lessons that will allow us to avoid repeating earlier mistakes, rather than only dwelling on current perceptions of what should have been.

3

THE ANTARCTIC COLLAPSE

UNCERTAINTY TAKES A (BRIEF) HOLIDAY

As discussed in chapter 2, the excessive catching and investment by the five Antarctic nations during the 1950s was largely the result of the postwar demand for fats, the intense competition encouraged by the single quota system, and—most importantly—the general unwillingness of governments to accept the Scientific Committee's majority view that the Antarctic quota was dangerously high. But by the 1958 IWC meeting, it was becoming increasingly clear to many in the commission that the so-called whaling olympic mode of hunting was driving both the industry and the fast-dwindling Antarctic stocks to ruin, leading the UK commissioner—with the strong support of Norway—to propose that the five Antarctic nations replace the competitive single quota system with preassigned national quotas.[1]

As had been the case since the IWC's inception, however, conservation measures within the commission at the end of the 1950s remained secondary to the economic and political interests of the Antarctic whaling industries, as defined by criterion I (utility) and criterion II (compatibility with already established needs).[2] The British and Norwegian push for national quotas essentially was an attempt to control the uneconomical and often chaotic way in which Antarctic whaling was being conducted. But in effect the scheme provided little in the way of protection for declining stocks, since the overall number of whales taken would remain mostly unchanged under the national quotas system. The SC's calls at earlier meetings for both reductions in the number of whales taken and also fundamental changes in the commission's management methodology—most significantly the abandonment of the Blue Whale Unit in favor of species-by-species quotas—were still no closer to being heard in the IWC.

Most commissioners continued to insist on what could be described as the "precautionary approach" to management: a focus on uncertainty to justify the interests of the industry over conservation measures that could not convincingly demonstrate their necessity.[3]

Thus, by the mid-1960s, the end game of Antarctic whaling was being played out to a conclusion made inevitable by fierce competition, itself caused by the single quota system and the Antarctic nations' unrelenting preoccupation with recovering the exaggerated investments in whaling of the 1950s. The issue of uncertainty no longer concerned, as it had during the fin whale debate, for example, the question of whether a reduction in the Antarctic quota was needed; the SC's conclusions, supported by the ever dwindling catch levels, were now strong enough to exclude any alternative explanations. The commission's attention instead shifted to the question of how far reductions in the quota should be taken. This change in focus, however, did little to reduce resistance to conservation initiatives of those planning either to continue operations in the Antarctic, which by 1965 was only Japan and the Soviet Union, or those who needed catching to continue at least until they had cut their losses by selling their interests and leaving the industry, as was the case with the United Kingdom and the Netherlands.

For the United Kingdom, the Netherlands, and later Norway, the utility of whaling had changed fundamentally by 1965 due to both the collapse of the Antarctic stocks and the increasing availability of other sources of edible oil and fats. The Dutch and British industries recognized that there was no longer any utility in continuing whaling operations because of the losses they were suffering. Their perspectives on whaling (in terms of criterion I, and also criterion II since demand for whale products was in decline by this stage) became completely different from those of Japan and the Soviet Union, where whaling and its products were still financially and politically viable, relevant, and in demand. Thus, the British and Dutch companies pulled out because the circumstances of the 1960s meant more utility in selling their respective interests and ceasing whaling than in competing for an increasingly scarce resource for which demand was steadily dwindling; whaling for the British and Dutch was no longer relevant to any established need as per the criterion II definition.

For Norway too, there remained little point in maintaining a presence in the Antarctic. The economic rationalization that the Norwegian government had hoped would accompany the national quotas scheme never materialized due to the long delay in the scheme's implementation, and also the effect this delay had in accelerating the collapse of Antarctic whaling. But an important difference between Norway's circumstances and those of the British and Dutch industries was the long association and strong cultural links with whaling. This caused the shift in perceived utility only to change, rather than end, Norwegian whaling as we will see in the following chapters.

This chapter's discussion of the crisis sparked by the national quotas issue and the changes that followed in the way the IWC treated scientific advice will show how the use of scientific uncertainty as a justification for delaying conservation measures by the commission became increasingly difficult from 1959 onwards. The apparent empirical "reality" that there were in fact fewer whales to be caught in spite of increasing effort and more efficient catcher boats[4]—in addition to the SC's improving ability to convincingly account for many of the uncertainty issues that had so far undermined its calls for lower quotas (due to the commission's eventual enlistment of external population dynamics experts in 1960)—gave scientific advice a stronger voice and made it more difficult for warnings of stock depletion to be ignored on the basis of scientific uncertainty.

Despite the dire situation confronting the commission at the end of the 1950s, the increasing evidence of overhunting still was not sufficient to force more conservative policies from all in the IWC. The growing scarcity of fin, humpback, and blue whales, however, did make it much more difficult to dispute the conclusion that these species had indeed been overhunted. Those arguing against lower quotas were forced to more openly dismiss the potential consequences of excessive hunting (as per the warnings of the SC) as a lesser concern than the looming specter of financial losses, indicating that for some governments in the IWC the danger of a species disappearing altogether was a more acceptable outcome than the threat of a smaller Antarctic quota causing financial ruin.

In illustrating how the nature of scientific uncertainty and its application within the IWC changed, this chapter will explore specific developments within the commission that might tell us how political and economic interests were influencing the commission's treatment of scientific advice:

(a) the withdrawal of Norway and the Netherlands from the IWC due to the failure of the five Antarctic nations to agree on national quotas (Japan had also signaled its intention to leave, but then chose to remain in the commission), and the resulting absence of any Antarc-

tic quota between 1959 and 1962, with catches exceeding 17,000 BWU in each of those seasons;
(b) the IWC's appointment in 1960 of a Committee of Three (three independent scientists) in an attempt to reestablish quotas by finally resolving the fin whale and Antarctic quota debate; and
(c) Japan and the Soviet Union's treatment of scientific advice in relation to their newly acquired status as the two biggest pelagic whaling nations, following the collapse of the major Antarctic stocks by the mid-1960s, the subsequent withdrawal of the United Kingdom and the Netherlands from pelagic whaling, and the decline of Norway as a whaling power.

THE NATIONAL QUOTAS DILEMMA: DIVIDING AN IMAGINARY PIE

In terms of reducing the chaotic competition engendered by the single quota approach, and relieving the financial strain of continued investment in infrastructure, the adoption of national quotas certainly had merit. One important problem with this initiative, however, was that not all the Antarctic nations regarded separate quotas as being in their best interests. The economic pressures of whaling were having the most immediate impact on the privately funded British and Norwegian companies, which did not benefit from either state support, as was the case with Dutch and Soviet whaling, or the existence of a strong domestic market for whale meat, as was the case in Japan. But while the Japanese and Soviet governments were not particularly concerned about economic issues, since the former was making money and the latter did not need to, the state-supported Netherlands Whaling Company was experiencing mounting political pressure at home by the end of the 1950s to become self-sustaining. And as we will shortly see, the high quota demanded by the Netherlands for the company's single Antarctic operation became a major obstacle to agreement on how to divide individual shares within the total quota set by the IWC.

The pressures resulting from what each industry and government regarded as an acceptable quota share raised a fundamental (and familiar) problem: that agreement on national quotas would also require agreement on a total Antarctic quota—something that had so far eluded the IWC in general, and the five Antarctic nations in particular. The adoption of a national quotas scheme, therefore, was ultimately dependent on the unlikely prospect that some consensus concerning an overall quota could be reached: a task made virtually impossible in the commission by the prevalence of "uncertainty" in the advice of the IWC scientists, and the willingness of various governments, in particular the Dutch, to use the element of uncertainty to pursue solely economic interests. Quota negotiations between the five governments were further hampered by the ICRW itself, because the charter did not allow for individual national quotas or for limiting factory ship numbers.[5] Thus, any discussions and agreement on this issue had to be pursued outside the IWC, which further complicated proceedings by bringing the convention's relevance to whaling regulation into question: if the Antarctic nations felt it necessary to set quotas and other limits on hunting outside of the ICRW, what was the point of the convention in terms of regulating whaling?

The many difficulties that stalled a decision on national quotas, until agreement was finally reached in 1962, have been covered in detail elsewhere.[6] For the purposes of this study, the most important question relating to these negotiations concerns how the Antarctic nations treated the growing evidence that there simply were not enough whales left to satisfy everyone, given what each government argued its industry needed, and also the role uncertainty played in justifying how big a slice of the "Antarctic pie" governments believed they could demand.

Negotiations on national quotas began soon after the London IWC meeting in June 1958, where the United Kingdom had proposed the scheme's adoption. The five Antarctic nations met at the London Whaling Conference of November 1958 to discuss the creation of a quota agreement intended to cover a seven-year period beginning with the 1959–60 season. After agreeing in principle to a national quotas scheme, the Antarctic nations then committed themselves to negotiations on the quotas (to be completed before June 1, 1959) and produced a short set of recommendations for the framework within which the discussions were to take place. The London conference stopped short of any agreement on what the actual quotas for each country would be, but it was decided that the Soviet Union would receive 20 percent of the total Antarctic quota and that the total quota should be the same as that set under the ICRW (the Antarctic quota having just returned to 15,000 units for the 1958–59 season after having been reduced to 14,500 for the previous two seasons).[7] The Dutch delegation, however, signaled its strong reservations on a total quota of 15,000 units, arguing that it would produce shares too small to be worth pursuing. According to Tønnessen and Johnsen, "the only way out," in the Dutch view, would be

> to leave the Commission and operate freely for three months. It was calculated that the catch would come to about 24,000 units. The scientists of every country—except the Dutch—agreed that within a very few years this would utterly decimate the whale stocks. It is not hard to imagine what the result of 24,000 units spread over five years would have been. In 1964–65 fifteen expeditions could only account for 7,000 in three and a half months.[8]

The Soviet government earlier had made clear its intention to further expand its Antarctic fleet in the near future (from one to four factory ships). For this reason the other governments accepted the Soviets' demands in the interest of keeping the negotiations going, even though this represented a share that was roughly three times what the Soviet fleet had averaged from the total Antarctic catch over the previous decade.[9] This agreement was based on the understanding that the Soviet Union would introduce no more than three new factory ships during the quota period and that the other Antarctic nations would not be able to add any additional factory ships to their operations, with the exception of ships already in operation in the Antarctic purchased from another of the five nations along with "a proportionate part of the quota of the vendor's country."[10]

Thus, the four remaining Antarctic nations were left to agree upon their respective shares of the Antarctic quota by the next IWC meeting in 1959—a daunting schedule given the obstacles to agreement and potential consequences of failure that already had arisen out of the London IWC meeting. In addition to the problem of finding a compromise on quota distribution that would satisfy each of the four remaining parties, further pressure was added shortly after the London conference by announcements of resignation from the IWC by Norway (December 29), followed by the Netherlands (December 31), and lastly Japan (February 6). These notices of withdrawal from the commission were each subject to cancellation by the governments, should agreement on national quotas be reached before the close of the 1959 IWC meeting.[11]

As mentioned earlier, the Norwegian and British industries were private enterprises run without any direct government financial assistance, and for this reason they clearly were the most enthusiastic about introducing a more economically rational approach to whaling. Norway, with its strong historic and economic ties to whaling (Norway had the largest fleet and accounted for more than 39 percent of the entire Antarctic catch between 1947 and 1958), saw the national quotas scheme as "the only salvation" for Norwegian whaling in the long term and on these grounds the Norwegian government made clear its intention to resign from the IWC if negotiations failed.[12] The United Kingdom's whaling interests were considerably less than those of Norway both in catch and infrastructure, and for this reason, perhaps, the British were less inclined to view resignation from the commission as a last resort should their proposal fail. Nevertheless, the British government clearly was equally concerned with the impact the whaling olympic mode of hunting was having on the current, and future, profitability of Antarctic hunting. Thus, with both Norway and the United Kingdom appearing strongly committed to the national quotas scheme, and the Soviets appearing satisfied with the offer of 20 percent, all that remained was to ensure the participation of Japan and the Netherlands. But this would prove to be no easy task.

The Japanese delegation was less enthusiastic than either the British or Norwegians about the national quotas system. The Japanese representatives continually referred to the legal problems involved with setting up a regime that violated the ICRW, but were placated to a large extent by a provision allowing for the purchase of factory ships and quotas. However, the consensus among many who observed or have examined the dispute over national quotas seems to be that the Dutch were by far the biggest obstacle to an agreement, due to their demands for a larger quota and their insistence that the overall Antarctic quota could be increased. According to Elliot:

> The Russians, much to everyone's surprise, quickly agreed both to the principle and to a quota for themselves of 20 percent—far higher than justified for their one expedition but perhaps not excessive for four.... [Norway and the United Kingdom] were well aware that the industry was on the edge of an abyss, and that only a national quota agreement gave some chance of stability and of *the drastic reduction in overall catch* that was required. The Japanese were less convinced that there was any danger; they were doing well themselves and were in no hurry to change the regime. The Dutch were the real stumbling block. Not only did they refuse to accept any evidence of stock decline, but they also stood out for a national quota far higher than could be justified on any basis.[13]

Important to any discussion of the national quotas dispute is, as Elliot has pointed out, the question of the total Antarctic quota. Dutch insistence on a much higher quota per expedition (i.e., one floating factory plus catchers) than that accepted by the other four nations made any agreement on a total quota unlikely. The only way to accommodate the Dutch demand for 1,200 BWU per expedition (the next highest was Japan's demand for 817 units[14]) was either to reduce one or more of the other quotas to make up the difference or to increase the total quota, which already had been raised from 14,500 BWU to 15,000 BWU at the 1958 IWC meeting (due to an objection lodged by the Netherlands citing insufficient evidence for maintaining the lower quota).[15]

The commission, since the 1953–54 season and in spite of Dutch arguments for quotas of 16,000 or more (16,500 including humpbacks), had managed at least some reductions—dropping from 16,000 BWU to the low of 14,500 in the 1956–58 seasons—even though these changes still were well short of the numbers called for by the SC. Elliot's assertion, however, that a national quota system provided "some chance of stability and of the drastic reduction in overall catch that was required" is highly questionable in view of all the Antarctic nations' willingness to reject the reductions recommended by the SC on the grounds that the evidence of decline was not strong enough to risk the interests of the whaling industry—a point made clear by the British commissioner at the 1955 meeting in Moscow.[16] The SC had warned, as recently as the 1958 meeting, that even "the present level of catching [i.e., the reduced quota of 14,500 BWU] is too high"—a view the majority of the commission accepted, but responded to by voting to keep the limit at 14,500 units with the Netherlands dissenting. The Dutch believed 16,000 units (not including humpback catches) was more appropriate and later objected, prompting additional objections from the other Antarctic nations and a subsequent return to 15,000 BWU for the 1958–59 season.[17]

At the 1959 meeting, just prior to Norway and the Netherlands effecting their resignations, the SC again warned of the declining fin and blue whale stocks—the Dutch scientist Slijper was the only scientist dissenting on the grounds of insufficient evidence—but no proposal for a lower quota was made in the commission.[18] Even as late as the national quotas system's proposal at the 1958 meeting, it was obvious that none of the other four Antarctic nations was prepared to go below 14,500 units, which in the majority view of the Scientific Sub-Committee (the exception was the Dutch scientist Drion) was still too high.[19] Furthermore, if one considers that all the governments involved in the national quotas negotiations (except the Soviets) were unable to agree on separate shares with 15,000 units as a total quota, it seems unlikely that any "drastic reduction" in the Antarctic quota could have been realized.

So while the Dutch position can easily be seen as unreasonable in terms of the actual number sought for its one expedition (the Dutch average for the four previous seasons was only 797 BWU), the rationale for their demand was essentially no different from that of any of the other four governments; the driving force behind each government's position was not better protection for stocks, but rather the interests of their respective industries. The Dutch, it seems, were simply being more excessive than the other four. As Tønnessen and Johnsen also have noted, the real issue at stake was the short-term profitability of whaling, and for this reason setting the quota at a lower level was never really a priority. Indeed, the setting of a lower quota, it appears, was always going to lose out to the greater need to catch what the industries believed was "enough" whales:

> [Speaking at the 1959 IWC meeting] the British representative referred to a government statement made in Parliament to the effect that if the agreement were not concluded and the Commission failed to save the whale stocks, it was doubtful whether Britain should continue as a member. But—and this boded ill for the agreement—Britain could not accept a lower quota than she had demanded [i.e., 2,250 BWU for three expeditions], as this was a minimum for profitable catching. The statement failed to reveal any really profound understanding of the situation; "a profitable quota" could hardly be a British privilege. It seemed to *presuppose that there were still sufficient whales available*; the problem was merely to put the quota so high that whaling would show a profit. Naturally, this was what everyone wanted, and for this reason it was impossible to conclude the agreement, as no one was willing to lower his sights.[20]

By the 1959 meeting in London, the IWC clearly was in crisis. Three conferences earlier that year (two in Tokyo and one in Oslo) had failed to bring agreement on national quotas any closer, and the commission now faced the prospect of having three of the five Antarctic whaling nations leave the commission to continue hunting outside the ICRW. The Norwegian commissioner stated at the meeting that it would cancel its resignation if a national quota agreement giving Norway 4,850 BWU from a total Antarctic quota of 15,000 units came into effect before June 30. But given the quota demands of the other three countries, in particular the Netherlands, there was little hope of Norway's ultimatum having any effect on the outcome—especially at this late stage. The Netherlands commissioner, for his part, informed the other delegates that the Netherlands Whaling Company would not exceed a catch of 1,200 BWU over the next seven years and would continue to adhere to the convention's 1958–59 schedule "with the exception of the restrictions concerning the number of whales to be caught and the period during which they might be taken."[21]

The Japanese government cancelled its resignation at the last moment, joining the United Kingdom and the Soviet Union as the only Antarctic nations still bound by the 15,000-unit quota set under the convention. But with Norway and the Netherlands now operating freely, there was little point in leaving the IWC; each country was now able to get, with interest, what they had been unable to obtain under the proposed national quotas system. The following numbers show the individual quotas each government gave itself at the 1959 meeting, with the national quota they had been prepared to accept in parentheses: the Netherlands 1,200 (1,200); the United Kingdom 2,500 (2,250); Norway 5,800 (4,850); Japan 5,100 (4,900); and the Soviet Union 3,000 (3,000). The unsanctioned quota for the 1959–60 season amounted to 17,600, but, rather ironically, only 15,512 units were taken (compared with 15,301 in the previous season) because the Norwegian, Dutch, and British fleets were unable to fill their self-appointed quotas.[22]

For most of the 1950s, uncertainty arguments had been used by the Antarctic whaling countries to argue against any significant reductions in the Antarctic quota. With the failure of the national quotas negotiations in 1959, however, the issue of lowering the total quota below 15,000 had become a moot point on purely economic grounds—even though all the Antarctic whaling governments, with the exception of the Netherlands, had accepted the majority SC opinion that blue and fin whale stocks (the mainstay of Antarctic whaling) were in decline. Only Slijper still disputed the committee's opinion in 1959 that "the balance of evidence indicated a decline in the stock of the most important species—the fin whale. The blue whale catch had fallen to a new low level."[23] The other commissioners only paid lip service to the increasingly urgent need for a lower quota that was soon made even clearer by the failure of all the fleets, bar one, to fill the individual quotas that each government had demanded for the 1959–60 season.

Put simply, the situation by the close of the 1959 meeting had become more divided and desperate. Four of the five Antarctic governments openly recognized the need for a lower quota (the United Kingdom, the Soviet Union, and Japan still adhered to the 15,000 BWU quota; Norway supported it in principle) but were not prepared to sacrifice catching in order to achieve it, while the Netherlands continued to claim that a higher, rather than lower, quota was possible—basing its position ostensibly on uncertainty arguments, raised by Slijper and Drion, concerning the data used to justify the need for a lower catch. Others, however, believed these uncertainty arguments were only a front, noting that the Dutch position on the total quota had less to do with science and more to do with increasing government pressure. Dutch whaling needed to become profitable in the face of growing domestic political opposition to the large amount of public money the Netherlands Whaling Company had so far absorbed.[24]

A major cause of the failure of the national quotas negotiations in 1959 was the same issue that had generated the (ongoing) discord over fin whale numbers: the Dutch refusal to recognize that any reduction in the total quota was needed. Norway and the Netherlands' departure from the IWC essentially represented the means by which these two governments had chosen to try to force a resolution of the ongoing fin whale debate. Norway gambled, mistakenly, that its threatened withdrawal from the commission would be enough to push the national quotas scheme through under a ceiling of 15,000 BWU. This strategy failed because the Dutch were prepared to risk the recovery of the remaining stocks by going it alone.

In view of the absence of irrefutable evidence that the blue and fin stocks were being pushed beyond the point of no return, which appears to be what the Dutch and, to a lesser degree, other nations were demanding, and also the high financial stakes involved in pelagic whaling in the Antarctic, it is perhaps hardly surprising that the commission eventually lost control over Antarctic whaling altogether by 1960. Gambell summarized the situation facing the commission:

> There was no precise figure that anybody could come up with and say this is the number of whales you can catch, and the whaling industry of course always wanted to catch the maximum number that it possibly could. And particularly the Netherlands, in the 1950s, early '60s period, were very anxious, with their one whaling operation, to get as many whales as they could—as much profit as they could—and so they argued very often that, in the absence of scientific knowledge, you should do what you think best.

And [the Netherlands] thought that catching more whales was best for them. Everyone else of course had to go along, because if one country decided to catch more, everybody needed to catch more in order to stay in place.[25]

The 1960 meeting in London was the first without Norway and the Netherlands, and its discussions, not surprisingly, focused mainly on the joint problem of resolving the quota dispute and enticing these two governments back into the commission. At this meeting, the UK commissioner took the lead in attempting to create a solution, while also suggesting that his government too might leave the IWC if the situation could not soon be rectified. The British attempt, however, was only a further step in the wrong direction, given the increasingly serious situation facing the Antarctic stocks of fin and blue whales, since it amounted to little more than "officially" suspending the Antarctic catch limit altogether for the 1960–61 and 1962–63 seasons—something that had already occurred in practice after the voluntary setting of quotas at the previous meeting. The reasoning behind this strategy basically was to buy enough time for agreement on the national quotas and on an Antarctic quota, because, in the United Kingdom's view,

> it appeared that these two countries [i.e., the Netherlands and Norway] had left the Convention because they considered it wasteful and uneconomic for the five Antarctic pelagic whaling countries to engage in competitive catching against each other within the blue whale unit ceiling laid down, and it seemed to the United Kingdom that satisfactory conditions for exploiting the stocks would not be secured without agreement between the Antarctic whaling countries on arrangements for sharing the total catch.[26]

The main impediment to Norway and the Netherlands returning to the commission was the question of what the Antarctic quota should be, since (a) there was not enough to satisfy everyone and give the Dutch 1,200 BWU within a limit of 15,000 BWU; and (b) the Norwegian government had already made it clear at the previous meeting that Norway would accept no higher an Antarctic quota than 15,000 BWU. But it is important to remember, and on this point the British proposal is somewhat confusing, that the Norwegian resignation appeared only to be based on disagreement over the division of the then Antarctic limit (i.e., 15,000 BWU), rather than on a demand for a higher total quota of units as had been made by the Dutch.

This point is supported by the statements made by Norway and the Netherlands after the 1960 meeting in response to the calls for their return to the commission. The Norwegian government rejoined the IWC in September 1960, in spite of strong protests from Norway's whaling companies,[27] but stated that its adherence to the convention depended upon the following three points: (a) the Netherlands' rejoining the commission; (b) the Soviet Union adhering to its earlier acceptance of 20 percent of the total Antarctic quota set under the ICRW; and (c) agreement on national quotas being reached within a "reasonable time." The Netherlands government, however, chose not to rejoin and informed the commission that its return "was conditional on prior agreement being reached on the future level of the total catch limit . . . and on the establishment of national quotas for the countries engaged in Antarctic pelagic whaling."[28] Thus, while Norway was prepared to rejoin in order to resolve the quota issue, the Netherlands refused to rejoin until the problem of a total quota and national quotas had been resolved.

The British proposal to suspend the Antarctic quota was accepted "with some reluctance" by a vote of seven in favor, two against, with four abstentions. Interestingly, Japan and the Soviet Union both voted against the proposal and lodged objections, leaving them as the only two IWC members still obligated to the 15,000 BWU quota.[29] But Japanese and Soviet objections were of no real significance (other than adding support to Norway's stance on the Antarctic quota), as the only limit now came from a request by the IWC for each government not to exceed the previous season's catches. Thus, the accepted 1960–61 limit for Antarctic pelagic catching was 17,780 BWU, slightly higher than the 1959–60 season (due to a small increase in catching per expedition by Japan), from which only 16,433 units were taken. And once again, only the Japanese ships managed to fill their quota (5,980 BWU).[30]

Apart from the small increase in the overall total, the other important difference in the 1960–61 season was the expedition numbers. The Soviets introduced another new factory ship, the second since the 1959–60 season, bringing their total to three. But more importantly perhaps, in terms of signaling a future trend that will be discussed later, the United Kingdom had sold one of its factory ships, the *Baleana* (along with an agreed quota of 700 BWU) to Japan after the 1959–60 season. This sale reduced the United Kingdom to two expeditions and raised Japan to seven, making the Japanese the second largest of the pelagic operators behind Norway, who had eight factory ships.[31]

It was by this stage very clear that agreement on a total Antarctic quota was essential if the stalemate between Norway and the Netherlands over national quotas was to be broken and Antarctic whaling was again to be brought under the control of the commission. With this in mind, and no doubt also the wrangling over uncertainty that so far had characterized the IWC's attempts at setting Antarctic quotas "based on scientific findings," the British commissioner also proposed, in addition to suspending the Antarctic quota, that

> a small committee of three scientists qualified in population dynamics or some other appropriate science should be appointed by the Commission to carry out an independent scientific assessment of the condition of the whale stocks in the Antarctic which would provide a scientific basis for the consideration of appropriate conservation measures by the commission.[32]

The UK proposal for "an independent scientific assessment" was a clear recognition of the uncertainty issues that had so far undermined the SC's advice, and also of the role these issues had played in bringing the commission to its current impasse. The reintroduction of an Antarctic quota under the convention was the only way to get both Norway and the Netherlands to return to the IWC, and the only way this could be achieved was by setting a quota based on scientific advice that everyone, in particular the Dutch, would accept as valid.

Prior to this period of crisis within the commission, each of the Antarctic nations had been prepared simply to treat uncertainty as a basis for maintaining the BWU quota at a limit that satisfied the needs of the industries. Now, for the first time in the entire history of whaling, catching on the basis of what was believed to be available, rather than simply what was needed, was being given priority. But scientific uncertainty remained the key issue, and the task of reducing it to acceptable proportions now was considered achievable through an independent assessment by recognized specialists. And on the basis of this assumption the commission adopted not only the British proposal for the appointment of what would become known as the Committee of Three, but also the provision that "the Antarctic catch limit should be brought into line with the scientific findings not later than 31st of July, 1964, having regard to the provisions of Article V (2) of the Convention."[33]

What the commission did not appear to be considering, however, was the caveat that uncertainty over stock numbers could not be removed entirely. Indeed, having treated uncertainty for so long as something that could be used to justify catching what one wanted, the Antarctic nations now suddenly appeared to believe that science could provide advice that was not complicated by any uncertainty. As a result, the fundamental issue of how the (unavoidable) existence of scientific uncertainty should be understood and treated in the course of formulating policy was still being ignored by the commission—an oversight with implications that later would become all too clear.

THE NEW SHERIFFS IN TOWN: THE CREATION OF THE COMMITTEE OF THREE

The decision to go ahead with the formation of the Committee of Three (COT) at the 1960 IWC meeting had the general support of the commission, but some reservations were expressed. New Zealand was skeptical that the Antarctic nations would treat the advice of the proposed COT any differently than they had the advice of the IWC's scientific committees. Japan was not convinced of the need for a new committee since, in Japan's opinion, the IWC scientists already had all the necessary data before them—a curious position indeed considering the extent to which scientific uncertainty had become an issue in the commission. The Australian government, no doubt mindful of its shore-based humpback whaling interests, expressed concern that the COT should not override the authority of the Technical Committee and the commission in the setting of quotas, as there were many issues that needed to be addressed under the convention.[34]

The Ad Hoc Scientific Committee (AHSC), formed at the 1959 meeting, also was given an important role in the commission's attempt to establish sustainable quotas. This committee met for the first time in May 1960, one month before the plenary of the twelfth meeting in London where, due to the absence of the Dutch scientists (Slijper had accepted an invitation to attend as an observer but could not because of illness),[35] it was able to unanimously endorse a paper by British scientist Laws. The paper concluded—on the basis of declining catch per catcher day figures and increasing numbers of younger and immature fin whales in the catches—that fin whale numbers were indeed dropping. After concurring that the "combined evidence leaves no room for doubt of a decline of the fin whale stock in the Antarctic," the AHSC members went on to warn that they were "convinced that the stock cannot withstand the present rate of catching and that the only remedy ... is a substantial reduction in the total catch together with continued biological examination of the condition of the stock from year to year."[36]

In order to further verify the now "unanimous" opinion of the IWC's scientists that current exploitation levels were unsustainable,[37] the commission, on the advice of the SC, instructed the AHSC to work in conjunction with the COT scientists and "carry out a detailed and specified program to improve the collection and interpretation of data including the use of the latest methods of studying animal populations."[38] Thus, it was the commission's intention to have the AHSC, with the assistance of the COT, apply the latest available methods in the newly emerging field of population dynamics to a larger and better organized body of data. In addition to the task of assisting the AHSC, the three scientists were asked to "report within one year of their appointment the sustainable yield of [Antarctic stocks] in the light of the evidence available and on any conservation measures that would increase this sustainable yield."[39]

By March 1961, Douglas Chapman from the University of Washington, Kay Radway Allen from New Zealand's Marine Department, and Sidney Holt, a British scientist working in the Fisheries Division of the UN Food and Agriculture Organization (FAO), had all been appointed to the COT (by the IWC's chairman in consultation with the vice chairman and the chairman of the SC).[40] The COT met for the first time in Rome in 1961, from late April to early May, as did the AHSC. In addition to its own separate meetings, the COT also joined the AHSC as observers, giving advice on the kinds of data required for assessing the condition of the Antarctic stocks.

These initial meetings were concerned with preparing to assess stocks and sustainable catch limits, as requested by the commission. The COT for its part noted that, in addition to the need for close liaison between the two committees and the collection of further data, much of the data was held by various "national groups" and was not properly tabulated and organized in a form "suitable for population assessment." Other points raised included financing for the "considerable number of computations" required for estimates, and the need for a large-scale whale marking program in order to obtain the necessary data on movements and migration for the critically important task of "defining the limits of each stock."[41] (The IWC had earlier cancelled its whale marking program at the 1960 meeting due to funding difficulties.)[42] The COT, which had elected Chapman as its chairman, outlined its role as follows:

> The function of the Special Committee [i.e., the COT] should be to co-operate with the *Ad Hoc* Committee by guiding the preparation of existing data in an appropriate form, recommending what additional data are required, devising appropriate methods of analysis to determine optimum yield and assisting in the preparation of an estimate by these methods and to supply to the Commission an independent opinion as to the nature and reliability of the results.[43]

The AHSC, which included Slijper, divided itself into four subcommittees, each charged with assessing and organizing the currently available data with regard to its application for population assessment by the COT. The relevant areas investigated by the subcommittees indicate the kinds of information the COT deemed necessary and included a review of available data, measurement of catching effort, age determination, and whale marking.[44] Two of these categories, measurement of catching effort and age determination, had been the focus of many of the uncertainty concerns that surrounded the SC's earlier recommendations, in particular the controversy generated in the fin whale debate. Their inclusion in the COT's examination indicates both the perceived importance of these data by the scientists and also their intention to resolve the uncertainty issues that had earlier complicated the scientific committees' advice.

In short, the Age Determination Sub-Committee, which included Slijper, reviewed the three current methods—analysis of ovaries, baleen plates, and ear plugs—and recommended counts of the laminations on ear plugs as the most reliable method of determining age. The subcommittee also pointed out that a "problem with the absolute time scale" still existed in that it remained unclear as to whether "one or two laminations are laid down each year." The subcommittee suggested "as a working hypothesis it [be] assumed that there are two per year. Pending further evidence, age data will be presented in terms of the number of laminations as well as the number of years."[45]

The subcommittee dealing with the measurement of fishing effort stated that the accuracy of the catch per unit of effort as "an index of the abundance of available whales" could be "improved by a closer estimate of the time spent in hunting whales," and also by finding "some means of calibrating the efficiency of the vessels [i.e., catcher boats] so as to allow for the generally increasing efficiency of units over the years." To these ends, the subcommittee proposed that the Bureau of International Whaling Statistics in Norway be asked to prepare "a catalogue of the particulars of all factories and catchers and such other data as may allow for calculation of the catching time and a measure of the catching efficiency of the vessels based on the technical features and their catches."[46]

The COT and AHSC initially faced two important problems. First, there was the huge job of identifying and bringing together all the relevant catch and biological data that had been collected by the whaling companies and various national agencies, before then organizing and tabulating the information for analysis using computers. The second and more difficult problem was the quibbling that went on within the commission over the question of who should pay for the assessment of stocks by the two committees.

At their first meeting in 1961 in Rome, the COT made clear the need for "the data to be available for a meeting to be held not later than the end of 1961," if even a preliminary report on the status of stocks was to be made before the 1962 IWC meeting.[47] But the huge volume of data involved—including the organization of information on the species, sex, location, and year of more than 800,000 whales on record[48]—in addition to the apparent unwillingness of the commission to pay the extra costs that the assessment of stocks would involve delayed the COT's second meeting until December 1962. Consequently, the COT's second interim report, which analyzed the data collected up to (but not including) the 1962–63 season, was not submitted to the commission until January 1963, some ten months later than the original date called for by the commission.

At the time of the COT's appointment, the IWC's seventeen member governments were paying the relatively small sum of £250 each toward an annual budget that only just managed to cover basic operating expenses with no allowance for scientific research. The cost of the COT's assessment was estimated at the 1961 meeting to be £7,600, which would have increased each government's contribution to £700 in order to cover both the cost of the assessment and the existing budget—not a huge amount, even by the standards of the day, but many members simply were not prepared to pay.[49] Many governments instead complained that since the work of the COT was going to mostly benefit the five Antarctic nations, it should be these members who paid for the added expense in amounts proportionate to their respective quota shares.[50]

Without going into the details of the Finance Committee's deliberations and the various figures discussed by the commission, the upshot of this predicament was that even with the consent of the other four Antarctic nations, full funding for the COT's work would be delayed until the quota dispute was resolved and the Netherlands government returned to the IWC, which it did not do until just prior to the following meeting in 1962. By this time, however, only Japan had paid its share of the newly created "extra-ordinary budget" and, as a result, preparation and analysis of the data requested by the COT was delayed until well after the 1962 meeting,[51] as the IWC's *Thirteenth Report* explains:

> Plans for the card punching of data by the Bureau of International Whaling Statistics and the processing of the data by electronic computer and for a further meeting of the Committee of Three Scientists with the Special *Ad Hoc* Scientific Committee to assess the results were, however, delayed because the expenditure could not be authorized until the Antarctic pelagic whaling countries were in a position to assure the necessary funds. . . . This position was not reached in time for the data to be processed and for a meeting of the scientists to be arranged before the Commission's Fourteenth Meeting in July 1962. It is now expected that a meeting of the Committee of Three with the Special *Ad Hoc* Scientific Committee will be held later in 1962 and that the final reports will be considered by the Commission at the Fifteenth Meeting in 1963.[52]

Thus, in the time between the COT's initial meeting in 1961 and the second joint meeting in Seattle in December 1962 (which produced the "Second Interim Report" one month later), the only direct contact between the COT and the IWC's scientists occurred at the AHSC's meeting in June 1962, just days before the beginning of the IWC's fourteenth meeting in London in early July. However, only one member of the COT, Holt, was able to attend this meeting. But even without the benefit of Holt, Allen, and Chapman's analysis of the data the two committees had spent so much effort preparing, the AHSC felt confident enough with the data and analytical tools it had at hand to paint what was by far the clearest, and gloomiest, picture of the situation in the Antarctic provided by scientists to date.

On the basis of information indicating a serious decline in the Antarctic catch, drawn from the average catch of BWU per catcher boat for each day of hunting (i.e., catch per catcher day or catch per unit of effort), the AHSC concluded, without dissent, that the average take of 15,000 BWU per season throughout the 1950s indeed had been unsustainable. Further, they set the highest sustainable quota for the Antarctic stocks at no more than 5,000–7,000 BWU if the stocks were to be allowed any chance of recovering to the point where "reasonable" levels of hunting could be maintained. The June 1962 report of the AHSC scientists summarized both the situation facing the IWC and their collective opinion on how it should be dealt with:

> With an annual catch over the period 1953 to 1962 of about 15,000 B.W.U., maintained by a continuous increase in the effort during that period, the catch per catcher day has declined from about 0.9 in the years 1953 to 1959 to about 0.5 in 1961/62, that is to barely half its level at the beginning of the period. This decline in catch per catcher day includes no correction for increased efficiency of the expeditions and therefore reflects a decline in the stocks of at least this amount. It is clear therefore that the catch of 15,000 B.W.U.'s was greater than could be sustained by the stocks in the 1950s, and a level closer to the 11,000 B.W.U. recommended by the scientific sub-committee in 1955 would have been necessary to maintain the stock (at the level of those years).
>
> However, 11,000 B.W.U.'s removed from the present stock is a much greater percentage harvest than it would have been when first proposed in 1955. If the proportional catch recommended seven years ago were to be proposed now, the allowable quota probably should not be greater than 5,000–7,000 B.W.U.'s. This quota would then have to be maintained for some years to enable the stock to build up to a level at which a reasonable sustained catch could be maintained. If such a quota is not introduced at once, and if the catch is held at the level of recent years, a further and probably accelerated decline in stock will occur.
>
> ... The Committee hope that the Seattle meeting [i.e., the second joint meeting between the COT and the AHSC then scheduled for late 1962] will have sufficient evidence before it to decide what is the desirable level of stock abundance, what is the sustainable yield that can be taken at that stock level, and what levels of catch need to be set by quota, and for how long, to permit the stock to build up to that level.
>
> However, whilst the above is valid in general, the Committee must again draw attention to the inadequacy of setting an overall limit for several species together. Experience has shown that within such an overall limit, operations can be concentrated excessively on individual species and stocks.[53]

When the IWC commissioners gathered for the 1962 meeting in London, they had before them strong and unequivocal vindication from the commission's scientists of the past warnings made by the majority in the SC and its subcommittees, and also a new and more urgent warning that the IWC's management approach needed to change drastically. As had previously been the case in the 1950s, no one was able to "prove" that the Antarctic stocks were in decline; the difference was that the empirical evidence of overhunting had reached the point where no one in the SC was able or willing to offer an alternative explanation of what it meant. The CPUE data now indicated a clear drop in catches, but only after the damage, it seemed, had already been done; the challenge facing the IWC was not only to prevent the current situation deteriorating further, but to also prevent it from being repeated in the future.

However, the commission had other issues in mind, namely the return of the Netherlands to the IWC and the successful implementation of the recently concluded national quotas agreement under a total Antarctic quota of 15,000 BWU. Consequently, the 1962 meeting would achieve little more than a return to business as usual in the commission—an accomplishment that the IWC's membership seemed to be more than satisfied with even though it must have been quite obvious, as it was to the AHSC and no doubt the COT, that this alone would amount to little more than playing one's fiddle while Rome continued to burn in the background.

THE CONCLUSION OF THE NATIONAL QUOTAS DISPUTE AND THE COMMITTEE OF THREE'S BITTER MEDICINE

Prior to the IWC's 1961 meeting in June, a series of negotiations aimed at ending the impasse over the national quotas scheme took place. Three meetings were held between four of the five Antarctic nations (the Soviet Union sent an observer to the first meeting), with Norway, the United Kingdom, Japan, and the Netherlands coming close to an agreement on the division of quotas by the final meeting in May. At the 1961 IWC meeting in London, the UK commissioner gave an account of the three meetings that had taken place outside of the commission, explaining that,

> agreement was very near on a formula for an allocation scheme which would give 33 percent of the total permitted quota to Japan, 32 percent to Norway, 9 percent to the United Kingdom and 6 percent to the Netherlands, with 20 percent being allocated to the USSR. In addition to these basic percentages, under a special arrangement, the Netherlands would also benefit by a bonus system depending on the catch of her expedition by a certain date during the Antarctic whaling season.[54]

But the issue of just how close the Antarctic nations were to reaching an agreement at this stage remained somewhat unclear, since the final agreement was not actually signed until one year later on the final day of the 1962 IWC meeting. Without going into the details of what was to become the final act of the long-running national quotas saga,[55] the following main points of this final phase of negotiations are of concern to this study. At the first of the three meetings held in London in February 1961, four of the five governments, with the Netherlands as the exception, accepted the following BWU quotas (with the respective number of expeditions operating in the 1960–61 season in parentheses): the United Kingdom 1,350 (2); Japan 4,950 (7); the Soviet Union 3,000 (3, with the fourth and final Soviet factory ship to follow in the 1961–62 season); and Norway 4,800 (8). This allocation made a total of 14,100 BWU, leaving 900 BWU for the Netherlands' single factory ship—the largest of all the shares on a per-expedition basis.[56] The Netherlands, however, rejected this arrangement and also a further offer of an additional 80 units that had to be taken within the IWC season. The Netherlands instead demanded 90 "bonus" units and the meeting subsequently ended in stalemate due to a difference of 10 BWU.[57]

The next two meetings focused on the Dutch government's demands for a bonus quota, and this problem was resolved by the third meeting, again in London, in May. At this meeting, the Dutch finally accepted a proposal allowing the Netherlands Whaling Company to catch an additional number of BWU up to a total of 70, depending on the number caught by its expedition under the main quota. But because this arrangement meant that the size of the bonus depended on certain minimum levels being caught during the regular season, the Netherlands then demanded that the season be lengthened by bringing the beginning of the Antarctic season forward from December 28 to December 12. At its 1961 meeting the commission agreed to this change but rejected another proposal from the Netherlands for its expedition to be allowed to carry over "from one season to the next the balance of any unused quota allocation."[58]

A fourth meeting was held between the Antarctic nations in October 1961, where it was generally expected that the negotiations would finally conclude. But again this was not to be the case, because the Soviet Union refused to sign the agreement until all five countries were again IWC members, while the Netherlands now insisted that all five countries must sign the agreement before it would rejoin the IWC. This deadlock resulted in Norway announcing its resignation for the second time two months later in December, since its conditions for rejoining in the first place still had not been met. The Norwegian government, however, made its resignation conditional upon the national quota negotiations being successfully concluded by July 1, 1962.[59]

Ironically, the factor that finally saved the national quotas negotiations from yet another deadlock, with two nations again operating outside the convention, was the 1961–62 season's low catch of only 15,253 BWU against a total collectively set by the Antarctic nations of 17,780. The 1961–62 catch was the lowest since the 1957–58 season, when the total quota was still only 14,500, in spite of a catching season that recently had been lengthened by sixteen days and was now twenty-six days longer than four years earlier. Of the five Antarctic nations, only the four Soviet fleets met (and exceeded) their quota, while the Dutch fleet put in the poorest showing by managing only slightly more than half (615) of the 1,200 units it had set for itself. The average catch per catcher day for all the Antarctic fleets stood at only 0.51 BWU, 25 percent less than the previous season and almost half the average of the previous five seasons.[60]

Thus, at this point, the Netherlands no doubt realized that there was nothing to be gained by not rejoining the commission and accepting the national quotas agreement (and the accompanying 15,000 BWU Antarctic quota) as it currently stood, since there was obviously little chance of its fleet exceeding the already secured 900 BWU plus "bonus." Furthermore, the Soviets had confirmed in April 1962, shortly after the close of the Antarctic season, their intention to accept the existing quota agreement once the Netherlands rejoined.[61] Consequently, the Netherlands announced in May 1962 its intention to rejoin the commission,[62] thereby bringing to an end not only the Antarctic quotas dispute and Dutch assertions that sufficient proof of the need for lower quotas did not exist, but also the funding difficulties that so far had delayed the work of the COT and the AHSC.

But the ending of the quota dispute would provide no relief for the Antarctic stocks of blue, fin, and humpback whales—now, by all accounts, in a state of serious depletion—as the implementation of the agreement merely meant an official return by all to the 15,000 BWU Antarctic quota. On the basis of the AHSC report completed in June, just prior to the 1962 IWC meeting in London, the SC informed the commission that "the stocks of the most important commercial species in the Antarctic were considered to give cause for great concern."[63] The SC described the situation facing the blue whale stocks as one where "nothing short of complete protection for a period of years would meet the needs of conservation." Fin whale stocks, in the committee's opinion, were in danger of "further and probably accelerated decline," while the state of the humpback stocks was similarly grave with the exception of one particular stock, for which the SC noted "there is some expectancy of regeneration . . . provided there is no substantial increase in the present catch."[64]

But in spite of the poor catch in the 1961–62 season and the increasingly dire warnings from the SC and AHSC, the only major change in the commission appeared to be the absence of any further attempts by governments to justify resisting quota reductions or complete protection of stocks on the grounds of uncertainty, since uncertainty over the *need* for a lower quota was no longer an issue. Indeed, even the Dutch representative in the AHSC, Slijper, now seemed unwilling to dispute the opinions of his colleagues—a point illustrated by the fact that the scientific committees' reports no longer included dissenting opinions over the question of whether or not stocks were declining; the *belief* that they were was now unanimous.

At the 1962 IWC meeting, rather than questioning the veracity of the advice it had received, the commission—with the Antarctic nations at the forefront—simply "noted" the AHSC and SC's warnings and deferred any decision on the Antarctic quota until the next meeting. The Chairman's Report explained the commission's reasoning:

> The Commission considered the position of the blue whale unit catch limit which stood at 15,000 for the next season in the absence of any special action. They noted the view of the Scientific Committee that the stocks were in a bad way, that it would be desirable to obtain separate limits for special stocks and that the results of the special investigation were awaited. In these circumstances it was decided that the blue whale unit limit should remain as it was but it was hoped that the special investigations would be proceeded with as soon as possible so that the situation could be effectively re-examined at the next meeting of the Commission. The Commissioner for New Zealand expressed regret that the Commission was not inclined to take any action on this matter at the present meeting.[65]

But as irresponsible as this lack of action by the commission's members may seem in retrospect, their unwillingness to act despite the mounting evidence is hardly surprising. After three seasons without any limit on catching under the convention and several meetings without two of the

Antarctic nations as members, no one in the IWC was willing to risk any further discord by raising the issue of a lower quota than the 15,000 BWU limit that had only just come back into effect. Thus, the IWC's response simply was to avoid any further political aggravation within its ranks by doing nothing.

In any case, the commission's members had not obliged themselves to implement any further changes regarding the Antarctic quota, or the protection of particular stocks or species, until July 31, 1964, as per the IWC's 1960 resolution on the COT. Furthermore, sidestepping the need for reducing the total quota was also somewhat easier to justify because the COT, lacking funding from the commission, could not produce its second interim report (and, therefore, any specific recommendations for stocks) before the 1962 IWC meeting. Thus, in the interests of allowing some measure of political stability to return to the IWC, its membership opted for the "easy way out" by adopting a wait and see approach.

The only conservation-related decisions the commission was prepared to make at this meeting were (a) to agree that "as the condition of the baleen whale stocks was judged to be so poor, no extension of the open season was permissible, and that it should therefore remain from December 12 to April 7 for the next season [i.e., 1962–63]";[66] and (b) to call on Japan, Norway, the United Kingdom, and the Soviet Union to reconsider their objections to the schedule amendments, adopted two years earlier at the twelfth meeting, which shortened the blue whale and humpback season in the Antarctic (with complete protection for humpbacks in one particular area).

This was rather an empty gesture, because the commission then effectively invited the Netherlands also to lodge an objection by voting to allow the Dutch commissioner the opportunity to make an objection within ninety days of his government's return to the IWC. This offer was made on the grounds that the Netherlands' recent absence from the commission had prevented him from doing so earlier.[67] Not surprisingly, the Netherlands took advantage of the offer and subsequently lodged an objection.[68]

Two other decisions made at the 1962 meeting also deserve mention. The first is the IWC decision to abolish the AHSC on the grounds that the commissioners "felt that their work was being hindered and impeded as a result of the procedure by which the *Ad Hoc* Scientific Committee met before the meeting and reported to the Scientific Committee which met during the meeting after dealing with matters which should later be considered by the Technical Committee."[69] It is more likely that it was not so much the AHSC's scheduling as its findings that were the real source of hindrance to some in the commission (particularly since the AHSC's work had largely been treated with indifference up to this point). The commission solved this "problem" by instructing the SC to form its own "working parties" during the course of the year and then to meet to review the findings and compile its report in the week before the commission met "in order that the evidence is available from the beginning of the week."[70]

The second decision of note from the 1962 meeting concerns a series of discussions between all the Antarctic whaling commissioners on the creation of an International Observer Scheme. This idea was first proposed by Norway at the 1955 IWC meeting in Moscow and involved placing independent observers on factory ships for the purpose of monitoring the size and nature of each expedition's catch.[71] The scheme's implementation, however, was delayed first by the ratification of a protocol by all the commission's members permitting alteration of the ICRW to allow for such a scheme (otherwise deemed *ultra vires* under the convention), and then by the Antarctic quotas dispute, which as we have seen ended with the 1962 meeting.

The reasons for the observer scheme's decade-plus delay and the circumstances that contributed to its eventual implementation will be explained in the following chapters. It is sufficient to note that the 1962 IWC meeting saw some progress on the scheme being agreed to—due to the Netherlands' return to the IWC at the conclusion of the national quotas dispute—with proposals being submitted by the United Kingdom, the Soviet Union, and the Netherlands that, according to the Chairman's Report, "gave rise to considerable discussion" but no agreement.[72] Thus, an entirely new set of negotiations was about to begin between the Antarctic nations on a topic that was of considerable import to the management of the Antarctic stocks. Indeed, at a time when blue, fin, and humpback populations were generally believed to have reached critical levels, there was still no way of verifying the accuracy of the catches reported by the whaling fleets to the Bureau of International Whaling Statistics—a situation with potentially grave consequences for some time to come.

With the COT's funding problems now resolved, the three scientists went ahead with the long-awaited second meeting in December 1962. Meeting jointly with the SC, and with British scientist John Gulland joining the COT for the first time, the two committees considered the results of the COT's analysis of the data up to and including the 1961–62 season. The report of this meeting, the "Special Committee of Three Scientists Second Interim Report" made available to the commission on January 14, 1963, noted that although the meeting's scheduling did not allow for completion of all the data analyses, or for the results to be checked in detail, its members nevertheless agreed that

> the general conclusions, both qualitative and quantitative, have become clear, and point to the need for action so drastic and of such urgency that the Committee of Three think it essential that the Commission should be given an immediate interim report so that national delegations should have ample opportunity to consider the implications of the proposals before the 1963 meeting of the Commission.[73]

Such was the level of concern within the COT that the committee's report went on to state that while the scientists were aware that the commission had already declared its intention to bring the Antarctic catch limit "into line with the scientific findings not later than 31st July 1964," they felt that

> the Commission should be fully aware of the serious effects of continuing pelagic whaling at the present level up to and including the 1963/64 season, so that it may give consideration to the possibility of earlier action.[74]

As noted earlier, the work of the COT and its application of recently developed population dynamics methodologies represented a significant change in how scientists approached research on whale populations for management purposes. During the 1950s, the IWC's scientists had attempted to estimate mortality rates based on age determination methods (baleen plates and ovaries analysis, and later ear plugs), and they then used these rates as a basis of comparison with the mortality level they believed would allow for maintenance of a stable population. Using this approach, Laws and Ruud had determined by as early as 1955 that the mortality rate for Antarctic fin whale stocks had been too high over the previous six years and concluded, therefore, that fin stocks were in decline.[75]

This kind of methodology, however, was plagued by an ongoing controversy over the age determination methods, in which Slijper had played a leading role, that was not resolved until 1961 when the AHSC determined that counting the laminations on ear plugs was the most reliable method. But because this method had not been introduced until the mid-

1950s, there were still relatively few ear plugs available and certainly none had been taken prior to this time.[76] Problems also had been raised, again by the Dutch scientists, concerning the reliability of catch per catcher day data (or CPUE) due to factors such as weather, increasing competition between catchers, and increasing size and horsepower of catchers. But, as with the issue of age assessment, methods for dealing with these variables were established at the AHSC's 1961 meeting with the COT.

By more coherently organizing the large volume of data already available from the various agencies, such as the whaling companies and Norway's Bureau of International Whaling Statistics, and preparing it (for the first time) in a form suitable for computer analysis, the COT further improved the reliability of the critically important CPUE data (by taking the variables mentioned above into account). The COT's efforts, however, also provided a solution to the problems caused by the lack of ear plugs in mortality estimates by combining the available data from the (now expanded) whale marking program with age-length keys developed by scientists Beverton and Holt.

Analyses of the general decline in CPUE and also estimates of "surplus" stock were conducted using methods developed by De Lury and Schaefer respectively that had become widely accepted in fisheries management during the 1950s.[77] The Schaefer method, used by the COT to calculate sustainable yield for fin and blue whale stocks, had further developed the idea of "optimum fishing," itself developed by Norwegian scientist Johan Hjort during the 1930s (Hjort had used the concept in his work with the International Council for the Exploration of the Sea's Whaling Committee).

Schaefer's method of analysis for calculating maximum sustainable yield (MSY: the maximum average quantity of animals that can be taken regularly from a stock over time without causing that stock to decrease) represented one of the first major developments in modern fisheries management because it defined not only the notion of "overfishing" through its concept of MSY, but also provided a relatively simple means of calculating what the MSY for a given stock might be. Such calculations for whale stocks had so far proved elusive due to the lack of both reliable data and also expertise in population dynamics methodologies within the SC.[78]

The COT's 1963 "Second Interim Report" summarized the committee's assessment approach and succinctly explained the theory of fishing employed by the COT in making its recommendations to the IWC. This summary, quoted below, was soon followed by an almost identical summary in the COT's "Final Report," which also was available at the commission's 1963 meeting in London.

> Data for catches, catching effort, sizes and ages of whales, maturity and pregnancy and marking were analysed by a number of methods to determine the sizes of stocks and the changes in these, and rates of mortality and reproduction. An attempt was made to use all relevant data and to take account of present knowledge of the migration and division of stocks. Since most of the analyses necessarily relied rather heavily on measures of catch per unit effort as indices of the abundance of the exploited stocks, calculations were made of the effects of weather on catcher efficiency and of the increases in catcher efficiency over the years associated with increases in the size and power of catchers.
>
> In the early years of whaling the number of whales caught is small in relation to the total number in the stock, but as exploitation continues and intensifies, this proportion rises and the effect of exploitation in reducing the stock numbers becomes evident in a change in the catch per unit effort. This decline in stock is not in itself evidence of over-exploitation, though it does indicate that whaling is becoming a major factor determining natural stock size. Whales in a stock of a particular size and composition have a certain capacity for reproduction and a certain rate of mortality. The difference between these, that is, the excess of reproduction and subsequent recruitment to the exploitable stock, over the natural deaths is the measure of the "surplus" population, or in other words the catch which could be taken from that stock without either causing it to decline or allowing it to grow. This is what in this report we term the *sustainable yield* at any given time. The sustainable yield is zero in an unexploited stock. In such a stock which is neither growing nor decreasing effective reproduction and natural deaths must be balancing each other. As the stock is reduced by whaling the rate of recruitment, r (defined as the ratio of effective reproduction to stock size), must increase or the rate of natural mortality, M (the ratio of natural deaths to stock size), must decrease or both; this process results in an excess of recruitment over natural mortality which can be taken as sustainable catch.
>
> If we define the fishing mortality rate, F, as the ratio of catch, C, to total stock, P, then in any season, C = FP. But the fishing mortality to give the sustainable catch is Fs = r–M and Cs = P([r-]M). If in any season F exceeds Fs, the catch will be greater than the catch a stock of that size can sustain and the stock will decline. If the stock does decline, only an increasing amount

of effort, to increase F proportionately, will result in the previous catch level being maintained, and that only at the expense of a still further reduced stock. When the stock is very small the sustainable catch is also very small. At some intermediate size the sustainable catch is at maximum. Our analyses have been essentially to estimate values of F, r, M and P and to use these to determine for each stock: (a) the catch which the stock could at its 1961/62 level sustain without increasing or decreasing, (b) the maximum sustainable yield it might produce if permitted to rebuild, (c) the minimum time it would take for the stock to grow to the size at which it could sustain the maximum yield. (This would occur if whaling were temporarily suspended. If whaling continued, though at a reduced rate, such that catches each season were kept below the levels sustainable at that time, stock could eventually attain the optimum size, but this would take longer.)

The precision and range of data required to make assessment (b) are greater than those to make assessment (a), but as the optimum level is approached it may be estimated with progressively increasing confidence.... Since this is an interim report based on partial analyses, the estimates are subject to revision. *If they do err, it is most likely to be on the optimistic side, i.e. final estimates may well be lower.*[79]

The COT's findings, which confirmed those of the AHSC's earlier meeting in 1962 and were themselves reaffirmed and explained in more detail in the COT's "Final Report," warned that "sustainable yields of blues and humpbacks of any appreciable magnitude could only be obtained by complete cessation of catching of these species for a considerable number of years." The "Second Interim Report" also further supported fears that the fin stocks—upon which Antarctic whaling now heavily relied—were also depleted, stating that in order to avoid "further decline of the fin whale stock and allow it to build up to levels which will sustain high yields, the catches in the next few years must be drastically reduced to something less than 9,000 [i.e., 9,000 animals or 4,500 BWU]."[80]

The COT's "Final Report" and "Supplementary Report," supported by the "Scientific Committee Report," presented even worse news for the IWC—news that was unlikely to have surprised anyone in the commission by this point. Having assembled what was undoubtedly the most comprehensive compilation of catch and biological data, as the basis for the most detailed and best quantified estimates of stocks and recommendations for sustainable catch quotas ever presented by any group of scientists to the IWC, the COT's "Final Report" informed the commission that, in the unanimous opinion of the committee, there was no longer any justifiable doubt of the "serious danger of extermination" currently facing blue whales and some humpback populations, or of the fact that fin whales were "far below the levels of maximum sustainable yield."[81]

According to the COT, blue whales throughout the Antarctic and humpback whales in Areas IV and V of the Antarctic (two of the six IWC designated Antarctic management areas, which included the western Australian and eastern Australian/New Zealand fisheries respectively) required immediate and complete protection if the threat of extinction was to be avoided, and for at least fifty years if MSY levels were to be reached. The estimated catch of 9,000 or fewer Antarctic fin whales previously recommended in the "Second Interim Report" was downgraded to 7,000 animals (3,500 BWU)—a level that, if adopted immediately, would need to be maintained for "five years or more" in order for fin whale stocks to reach populations levels able to support a MSY.[82]

However, the "Supplementary Report," which included new calculations using information from the 1962–63 season (the previous reports went up to the 1961–62 season's catches), went even further, stating that on the basis of the most recent catch data the fin whale stock can only "provide a sustainable yield of about 4,800 whales [2,400 BWU]."[83] The SC, meanwhile, in its report to the commission and with the support of the COT, recommended "that the total protection of humpbacks should be not only in Groups IV and V, but in the whole Southern Hemisphere."[84]

The COT scientists also repeated the recommendation they had made in their "Second Interim Report" for the elimination of the BWU and the establishment of separate quotas for each species. Indeed, the IWC's ongoing preference for the BWU, despite calls as early as 1955 by the Scientific Sub-Committee for quotas to be managed on a species-by-species basis,[85] was a good illustration of the commission's simplistic approach to management, and also its generally low regard for the concept of MSY and the long-term benefits it might offer in terms of providing stable, sustainable catch quotas. At this juncture, with the main Antarctic stocks generally accepted to be at dangerous levels, the urgent need for more selective regulation of catching was obvious, as was made clear by the COT's reports to the commission:

> While the concept of the "blue whale unit" may have had some administrative convenience and given some apparent flexibility to the operation

of the quota system in the past, the maximum sustainable yield can only eventually be obtained by taking the maximum sustainable yield for each of these species [i.e., blue, fin, and humpback] considered separately. If, even after stocks of all species had built up to the optimum level, the total catch taken in a certain season were equal to the maximum sustainable total catch, but the species were not taken in the correct proportions, then the stocks of one or two of them would be reduced below the optimum level and the others would be allowed to rise above the optimum, with a consequent reduction in the sustainable catch of each and all of them. While, as at present, all stocks are below the optimum level, any incorrect distribution of the current total sustainable yield would involve over-exploitation of certain species leading to further reduction of their stocks or, at best, to deferment of recovery to the optimum level.[86]

The veracity of the COT's recommendations and warnings, compiled during the 1962–63 season, was further supported by another sharp drop in the total Antarctic catch. In the 1962–63 season, with the conclusion of the national quotas negotiations, the total Antarctic quota had returned to 15,000 BWU. But the total catch was only 11,306 BWU, as opposed to 15,253 units in the previous season, with a drop of approximately 30 percent in the fin whale catch (7,770) and 16 percent (171) for the number of blue whales taken.[87] And while only seventeen expeditions operated in this season, compared with twenty-one in the 1961–62 season, this made little difference since all but one of the fleets (the Japanese) were unable to fill their respective quotas or approach anything resembling their normal working capacity. The four Soviet expeditions came close to filling their quota; but for the Dutch, British, and Norwegian fleets, the season was an unmitigated disaster.

Before the 1962–63 season, the sale of two factory ships to Japan—one from Norway and one from the United Kingdom—caused the previously agreed upon quota arrangements to be adjusted in August 1962 (the sales involved a transfer of the quotas pertaining to each ship). Japan's purchase of the *Kosmos III* from Norway in 1961 and the United Kingdom's *Southern Venturer* in 1962 gave the Japanese fleet, which remained at seven expeditions, a total Antarctic quota of 6,150 BWU (or 41 percent). The United Kingdom was reduced to 750 units (5 percent) for its one remaining factory ship, and Norway to 4,200 units (28 percent) for its four remaining expeditions (three other Norwegian factory ships were also "laid up").[88] The Soviet and Dutch quotas and fleets remained unchanged. As noted above, the Japanese filled their quota, and the Soviets came up only 184 units short with 2,816 BWU, but Norway managed only 1,381 units, the United Kingdom 502 units, and the Netherlands a meager 457 units.[89]

But even with this evidence of whale stocks depletion before them at the 1963 meeting, along with the COT's dire assessments, the commissioners again were unable to act on the scientific advice they had received. And as at the previous meeting in 1962, there was no longer any pretension by any of the Antarctic nations that sufficient uncertainty existed to support alternative interpretations of the empirical evidence at hand.

The reason for the commission's failure to act was simply that the immediate economic and political costs of drastically reducing the quota, in the eyes of three of the five Antarctic nations (the Netherlands, Japan, and the Soviet Union), were just too great to be justified by any long-term economic or political benefits that may have occurred from allowing the stocks to recover. The Japanese and the Soviet fleets were still mostly filling their quotas, so neither government saw any utility in reducing the quota to a level where the recent expansions of their fleets would become meaningless at best and either financially crippling or (particularly in the Soviet case) politically embarrassing at worst. For the Netherlands, which seemed to be rather desperately holding out for a miraculous improvement in its catch, the cost was likely to be both financially and politically ruinous.

In the case of Norway and the United Kingdom, however, where whaling companies always had needed to be profitable based on whale oil sales alone, having no sufficiently large market for whale meat, there was no utility in continuing to hope that they could somehow realize profits from an industry that was literally dying. Indeed, the sales of Norwegian and British factory ships to Japan, in addition to Norway's laying up nearly half its fleet, already had indicated as much.

In the Technical Committee, following lengthy discussions on the reports of the COT and SC, BWU limits of 4,000, 10,000 and even 12,000 were proposed without any agreement. When presented to the commission, these proposals received an equally mixed reception—although only the 4,000 and 10,000 BWU limits were discussed in the plenary. During a lengthy debate in the plenary, Japan and the Soviet Union argued in favor of the 10,000 BWU ceiling, citing the great hardship their industries would suffer if a 4,000-unit quota was adopted. They were supported by the Netherlands commissioner, who justified his position by saying 4,000 units was not practical and adding that 10,000 units represented a significant reduction from the current limit of 15,000 BWU.

Norway, however, pointed out that since the 1963–64 season's catch hardly could be expected to be any higher than 8,000–9,000 units, a reduction to 10,000 represented no reduction at all. The United States supported Norway's position, while rather astutely noting that the arguments put forward for the 10,000-unit limit had nothing to do with conservation and were only based on the whaling industry's economic needs. The United Kingdom supported a limit of 4,000 units, but then agreed to accept the 10,000-unit limit for one season on the grounds that an adoption of 4,000 units would only result in objections from Japan, the Soviet Union, and, no doubt, the Netherlands. The Chairman's Report for the 1963 meeting summarized these discussions and their outcome:

> Eventually, on the proposal of the Commissioner for Japan and seconded by the Commissioner for the U.S.S.R. a blue whale unit limit of 10,000 was carried. There were 7 votes in favour [including the United Kingdom and Australia], 1 against [New Zealand] and 5 abstentions [Canada, Denmark, the United States, Norway, and France]. Some of the Commissioners who would have voted for 4,000 units decided not to as there were three countries who considered their whaling fleet economics could not be supported at such a level of whaling and they might be expected to object within the statutory 90 days if such a proposition were carried. The blue whale unit [limit] would then revert to 15,000 units.[90]

Of the COT and SC's other recommendations, only the total ban on humpback catches south of the equator was adopted, although New Zealand and Australia did unsuccessfully attempt to amend this motion so that shore-based hunting in both countries could continue for at least one more season.[91] The COT's recommendation that the BWU be replaced by species-by-species quotas was also rejected. The commission "noted this recommendation but felt that, for the present, the blue whale unit limit was the only practical method that could be administered."[92] Even the recommendation for total protection of the blue whale throughout the Antarctic fell short, despite the 1962–63 blue whale catch having declined further, dropping to 947, and now being nearly half the number of animals taken in 1960–61.[93] The reasons that the proposal for complete protection of blue whales was watered down, despite general agreement within the commission concerning this species' need for protection, are relevant

JAPAN RESORTS TO HUNTING PYGMIES; THE BRITISH AND DUTCH DECIDE TO CALL IT A DAY

The five Antarctic nations, with the exception of the Netherlands, abandoned the use of uncertainty arguments as a means of opposing quota reductions after the increasingly poor catches from the 1960–61 season onward demonstrated convincingly that the SC and COT's warnings could no longer be argued against on scientific grounds. Even the Netherlands finally admitted that uncertainty over the need to reduce the Antarctic quota was no longer a justifiable basis for resisting quota reductions.

Thus, by 1962 all the Antarctic nations had resorted to pleading economic hardship as the grounds for their joint resistance to lowering catching to sustainable levels. But while the usefulness of uncertainty arguments had largely evaporated in the face of difficult to contest evidence, important problems relating to the IWC's perceptions and treatment of scientific uncertainty still existed. There was continued absence of any attempt within the commission to recognize the unavoidable existence of uncertainty in science and to formulate any means of dealing with it.

The IWC's failure to adopt total protection for the blue whale illustrates the consequences of this ongoing failure to deal with uncertainty. After almost a decade of uncertainty issues playing a key role in thwarting the adoption of conservation initiatives in the IWC, scientific uncertainty suddenly became a nonissue and was ignored by the commission, in particular by the Japanese delegation and scientists, when the Japanese commissioner tendered scientific evidence at the 1963 meeting in support of a proposal for the continued catching of one particular stock of Antarctic blue whales.

Since the 1959–60 season, Japan had been reporting catches of what its scientists believed to be a previously unrecognized subspecies of blue whale (although some catches probably occurred in prewar whaling[94]) referred to as the "pygmy-blue whale," which is smaller and a different color than the larger blue whales traditionally favored by whalers.[95] Catches of these animals by the Japanese fleets from a stock in the north of Area III and part of Area IV had been increasing with the growing scarcity of the blue whale.[96] Faced with the prospect of a total ban on all blue whales in the Antarctic at the 1963 meeting, Japan's commissioner proposed that the ban should exclude this stock because it was "predominately a stock of pygmy blue whales and . . . his government attached particular importance to a continuance of the catching of these whales at least for the next year."[97]

The Chairman's Report for the 1963 meeting noted, "There was general agreement that this species [i.e., the blue whale] should be given the complete protection recommended by the Commission's scientists."[98] Nevertheless, the Japanese commissioner insisted upon his country's need to exclude the pygmy blue whale from the ban. Japan's position resulted in the United Kingdom proposing, with the support of Japan not surprisingly, that complete protection be given to the blue whale in the Antarctic south of 40 degrees south latitude with the exception of the earlier mentioned area inhabited by the pygmy blue whale stock. This proposal was carried with no votes against and five abstentions, no doubt because the other members feared that Japan probably would object and, therefore, remain free of any obligation to reduce its blue whale catch.[99]

The scientific advice concerning the pygmy blue whale stock at this stage was very sketchy, with little available information on the number of animals due to the very recent nature of the catches. And although the pygmy blue whale now is generally recognized as a subspecies of blue whale, some scientists at the time held strong doubts that this was the case. U.S. scientist Remington Kellogg, for example, raised the possibility that these smaller whales, about which almost nothing was known, were simply younger whales that podded together in this area.[100]

The COT scientists, however, in their "Final Report" concentrated only on the numbers related to this stock, rather than taxonomic issues, and warned the commission that the 1960–61 catch of 1,126 whales taken by the Japanese (the catch fell to 403 in the following season) had "greatly reduced the stock," which the COT believed had numbered less than 2,000 at the beginning of the season.[101] Then, in their "Supplementary Report," the COT revised their population estimates upward: based on more complete analysis of the "scanty existing data" by Japanese scientist T. Doi, in which he used catch information from the 1962–63 season, the COT thought the initial population probably had been higher and that the current number was likely to be somewhere between 2,000 and 3,000. But despite this revision, the COT emphasized that the sustainable catch was only 300–400 animals, and was more likely to be only half this estimate because of the strong indications that pygmy blue whales have a higher natural mortality rate than the larger animals. Thus, in the opinion of the COT, "the pygmy blue whale catch of about 700 in 1962–63 exceeded the estimated sustainable yield of this stock by a considerable margin."[102] In the 1963–64 season, the Japanese catch from this stock dropped to 113.[103]

In effect, no one had provided any generally accepted data on this stock, at least not until after the 1963–64 season when Doi and fellow scientist T. Ichihara submitted to the COT new stock size estimates, recommending that "the stock of pygmy blue whales could stand the annual catch of 400 whales for three seasons." The COT expressed "reservations" about Doi and Ichihara's stock and optimum level estimates but were in "general agreement" with their conclusion concerning the sustainable annual catch.[104] However, despite the generally recognized need for complete protection of blue whales and also the COT opinion that the data for the pygmy blue whale stock at the time of the 1963 meeting was "scanty," Japan—with the commission's support—still was able to go ahead with this catch citing nothing more than a "need" to do so and a body of "scanty" data that suggested a small catch may be possible.

It is important to note at this point that while scientific uncertainty is usually invoked in order to justify policies that are at odds with the scientific advice at hand, the obvious flip side of this relationship is for policy makers to ignore or downplay scientific uncertainty when its invocation is likely to undermine research they have provided in support of a given policy. So, faced with an absence of any significant levels of uncertainty concerning the need to protect blue whales in the Antarctic, Japan's strategy was to justify its proposed catch by arguing on the basis of both its national interest and also some generally weak estimates of the pygmy blue whale stock supplied by Doi.

Thus, the commission, with Japan at the forefront in this respect, still was treating uncertainty in a way that allowed whaling to continue; the only difference on this occasion being an indifference to uncertainty, rather than its invocation, as the means by which further catches were justified. One argument that could be made here is that if the data on this stock were so weak, the COT should have recommended against the catch on this basis, especially given the serious threat facing the Antarctic blue whales at the time. However, had they done so, the most likely result would have been for Japan to argue against this position by invoking uncertainty, since the "scanty" nature of the empirical data that were available for the stock also meant there was no way of convincingly demonstrating that the Japanese estimates were wrong.

A return to uncertainty arguments is precisely what occurred at the 1964 meeting of the SC, when the results of a final report by what was now referred to as the Committee of Four (COF) were considered (the COF included British scientist Gulland and was commissioned at the 1963 IWC meeting). The COF, in addition to repeating the need for quotas based on individual species, recommended the reduction of fin and sei whale catches in the Antarctic to the lowest possible level below their estimated sustainable yields, and also the banning of all blue whale catches—including pygmy blue whales—in the Antarctic.

Of the nine nations represented in the committee, including all the Antarctic nations plus the United States, Canada, Australia, and France, all but two endorsed the COF's recommendations. These dissenting scientists were from the only two countries that had been expanding their Antarctic whaling interests: the Japanese scientist H. Omura and the Russian scientist V. A. Tverianovitch. Both these scientists questioned the reliability of the data used by the COF, with Tverianovitch even abstaining in the vote on the need for species-based quotas.[105]

In the plenary of the 1964 meeting, the Soviet commissioner added further support to Japan's case by arguing that the available evidence did not support a ban on catches from the pygmy blue whale stock, adding that these catches would help reduce the pressure on fin and sei whale stocks.[106] In any case, the Japanese and Soviet treatment of uncertainty here—like the Netherlands in the fin whale debate—should be seen as little more than justification for a policy they were going to pursue regardless of the advice on offer, especially since Japan and the Soviet Union's trump card was their ability to lodge objections even if the commission had voted against Japan's proposal.

The Japanese and Soviet responses to the ban on blue whale hunting, therefore, provide a good example of how the whaling nations were able to adapt their arguments in order to continue resisting scientific advice that went against their perceptions of utility, as per this study's criterion I definition. From the early 1950s until the early 1980s, when the need to deal with scientific uncertainty finally was specifically addressed by the IWC, whaling governments supported their policies by using one or more of the following three methods, depending on the strength and general level of acceptance of the scientific advice at hand: (a) the invocation of scientific uncertainty to argue against any reduction when the empirical evidence was weak, as occurred in the 1950s, or to argue against only the size of the reduction when the evidence was deemed sufficient to justify some further limiting of catches, as occurred at the 1964 meeting; (b) simply ignoring scientific advice and arguing purely economic and political reasons when evidence of stock depletion became much stronger, as was the case in 1963 when the quota was maintained at 15,000 BWU; (c) selectively applying scientific uncertainty by downplaying any uncertainty issues that might undermine research put forward in support of catches, as Japan did in 1963 in the course of attempting to further support its case for continued catching of pygmy blue whales.

Another point in connection to Japan's continued catching of blue whales in the 1963–64 season concerns the question of why the United Kingdom—by this stage operating only one expedition filling just half its quota—appeared to take such a prominent role by proposing, on Japan's behalf, that catching of the pygmy blue stock be allowed to continue. The most common answer, suggested or implied by Small, Birnie, and Elliot, is that the threat posed by Japan's ability to veto the ban on full protection for all Antarctic blue whales, vis-à-vis the ICRW's objection provision for members, meant that the other commissioners had little choice but to accept Japan's proposal if the other blue whale stocks were to be receive complete protection.

But there is another aspect of how the United Kingdom voted at the 1963 meeting to consider. It relates not only to the issue of pygmy blue whale catches and the IWC's adoption of 10,000 BWU as the Antarctic quota, but also to how the era of the five Antarctic nations ended with only Japan, the Soviet Union, and Norway continuing pelagic whaling in the Antarctic by the 1964–65 season, and also with Japan, by this stage, as the largest operator in terms of both the number of factory ships and quota size.

In August 1963, one month after the IWC meeting, the British government informed the commission that the United Kingdom's only remaining factory ship, the *Southern Harvester*, "had been sold to a Japanese whaling company and that in accordance with . . . the Regulation of Antarctic Pelagic Whaling [i.e., the national quotas scheme] . . . , the United Kingdom catch quota of 5 percent had been transferred to Japan."[107] Given the increasing difficulty of filling quotas by this stage (in the 1962–63 season only Japan had managed to do so), it is not unreasonable to assume that Japan also was becoming concerned that it would become more difficult for the Japanese fleets as well, especially since Japan was the only country continuing to expand its share of a shrinking Antarctic quota. Furthermore, Japan's purchases of factory ships were only expanding the size of the Japanese quota and not the size of the fleet, since the Japanese were seeking a larger share of the Antarctic quota—the real reason behind Japan's purchases of factory ships. Initially, this strategy no doubt was intended to increase the size of Japan's total catch, but after the 1962–63 season, when it must have become obvious that the total quota would be reduced, it seems likely that Japan purchased *Southern Harvester* only to compensate for the increasingly smaller Antarctic quota.

Thus, with the sale of the *Southern Harvester* and its 5 percent share, in addition to the earlier purchase of the *Balaena* and Norway's *Kosmos III*, Japan's share of the total Antarctic quota jumped from 33 percent of 15,000 BWU (or 4,950 units) when the agreement was signed in 1962, to 46 percent of only 10,000 BWU (4,600 units) by the time of the 1963–64 season, thereby providing a net loss of 300 BWU for Japan's seven expeditions.[108] This situation raises the question of whether the United Kingdom's support for Japan's position on pygmy blue whale catches, and also the adoption of the 10,000 BWU Antarctic quota, resulted from more than only the threat of Japan and the Soviet Union lodging objections. Another explanation concerns Britain's desire to cover its losses by selling its last remaining factory ship to Japan, since it is also possible, given the situation, that the sale of the *Southern Harvester* may have depended on a minimum level of BWU being adopted and Japan continuing to catch blue whales.

No recognition of this not unreasonable connection, and the possibility of it having influenced how the United Kingdom voted in 1963, has been made in the existing accounts of the 1963 meeting. But in their discussion of the 1964 meeting, where resistance against catch quota reductions and total protection for the blue whale continued, Tønnessen and Johnsen point out the existence of a remarkably similar situation. The Dutch sold their only factory ship, the *Willem Barendsz*, and its accompanying 6 percent share of the Antarctic quota to Japan shortly after the 1964 meeting in Sandjefjord in Norway:

> Negotiations for the purchase of the *Willem Barendsz* . . . had been broken off by the Japanese in anticipation of the size of the 1964–5 quota. If this were drastically reduced, the Japanese were not interested in buying, and the Netherlands would stand little chance of selling the floating fac-

tory or its quota. No one could be in any doubt that this was the reason why the Netherlands voted in favour of Japan's high quota proposal. It was, furthermore, right up to the very last, entirely in line with the consistent Dutch resistance to any lowering of the quota. In their own interests they placed an economically advantageous winding-up of their company above preservation of whale stocks. Once the Japanese had been given the desired . . . quota, the purchase of the *Willem Barendsz*'s quota followed almost automatically.[109]

The circumstances surrounding the sale of the *Willem Barendsz*, in addition to the commission's inability to honor its earlier pledge of bringing catch limits into line with scientific advice by the 1964 meeting, will be dealt with in detail in the following chapter. This meeting marked not only the beginning of Antarctic whaling's final phase, with only three of the Antarctic nations remaining in operation, but also the beginning of an era in which the way scientific uncertainty influenced policy making in the IWC changed dramatically.

The return of uncertainty arguments in the course of Japan and the Soviet Union's attempts to keep Antarctic whaling profitable, in the face of growing scientific evidence for fin and sei whale catches to be abandoned, demonstrated at least a partial return to past practice. However, the increasing political pressure from both within and outside the IWC for the commission to further reduce hunting was at the same time causing uncertainty arguments to gradually take on a new complexion, particularly among governments whose economic stake in whaling had either disappeared, or was fast decreasing.

Thus, chapter 4 will lead us into a period of the IWC's development where the political priorities and dynamics within the commission changed—dramatically altering the way in which many in the IWC perceived and treated scientific uncertainty in policy making and setting the stage for the later adoption of the current moratorium on commercial whaling in 1982. Chapter 4's analysis of this period, which stretches through to the early 1970s, will explain how and why these changes constituted both a watershed in the IWC's direction and management goals and also the foundation for many of the problems that divide the commission today.

4

THE WORM TURNS

THE IWC'S REINTERPRETATION OF UNCERTAINTY

Chapter 3 discussed the circumstances that led up to the collapse of the Antarctic stocks and also much of the industry that relied upon them, with particular emphasis on how the use of scientific uncertainty arguments declined as empirical evidence of overhunting became more and more convincing. With catches dropping on a yearly basis, and also the Committee of Three providing much more detail than ever before on the status of the Antarctic stocks, the need for a significant reduction of the Antarctic quota was, by 1963, no longer disputed in the IWC.

But this situation did not prevent some in the commission—most notably Japan, the Soviet Union, and to a lesser degree the Netherlands—from seeking to delay the adoption of the SC and COT's recommendations, which, in addition to slashing the Antarctic quota, included abandoning the BWU, adopting species-by-species quotas, and providing complete protection for all blue and humpback whale stocks.

These delays initially were justified by the need to avoid further controversy at the 1962 meeting, following the discord that earlier had erupted over the national quotas debacle. Then, at the 1963 meeting, further delays occurred and were attributed to the commission's 1960 commitment to bring the Antarctic catch limit "into line with the scientific findings [of the COT] not later than 31st of July, 1964"[1]—a deadline that enabled the Japanese, Soviet, and Dutch commissioners to resist the COT scientists' January 1963 recommendations for an immediate reduction in catches for another year. But as the 1964 meeting loomed, these governments—the only members, with the exception of Norway, still operating pelagic whaling fleets in the Antarctic—were running out of reasons for resisting drastic cuts in the Antarctic quota for the 1964–65 season. It was increasingly likely that the now named Committee of Four (COF) would reiterate its earlier findings and also provide further warnings for sei and pygmy blue whale stocks.

There did remain two ways of rationalizing further delays in the implementation of the COT's recommendations, neither of which were unfamiliar to the commission by this stage. The first was simply to plead that cuts of the magnitude recommended would ruin what remained of the Antarctic whaling industries, an often-heard argument when the Antarctic nations were confronted with advice claiming the need to catch fewer whales. The second involved resurrecting uncertainty arguments, but applying them in a different way than before. The strategy, in terms of using uncertainty to resist scientific advice calling for unwanted hunting limits, became not to dispute the necessity of reducing the quota (as had been the case for much of the 1950s), but rather to use uncertainty to dispute the size of the reductions needed. As was indicated in the final section of chapter 3, the recommendations put forward by the COF—in an additional report requested by the commission at the 1963 meeting and completed in June 1964—received a less than enthusiastic reception from the remaining Antarctic nations at the 1964 meeting. The recommendations were subsequently opposed using a carefully blended mix of arguments pointing to both the imminent threat of financial ruin and the uncertainty that (inevitably) existed over the extent to which catch reductions needed to be made.

Thus, the IWC's treatment of scientific uncertainty at the 1964 meeting marked the beginning of a change in perceptions among the commission's members as to how uncertainty should be understood and accounted for—hitherto, uncertainty had been invoked only to support industry interests when evidence of overhunting lacked detailed analysis or was subject to alternative interpretations. This transformation in attitude would, by the late 1970s, result in a complete about-face in terms of the role played by uncertainty in the creation of IWC policies.

The issue of most significance, in terms of sign-posting the onset of this attitude change, was the COF's very clear exposition on the problems and long-term costs involved with waiting for incontrovertible evidence that a stock was being depleted. The gist of the COF's position (supported by their colleagues in the SC) was that given the available data and the important role played by CPUE figures in determining stock numbers, unambiguous evidence of a stock being overhunted could only be obtained when the stock already had become depleted and the animals so scarce as to cause a significant drop in catches—a flaw that later led to the abandonment of CPUE data as a means of estimating population numbers.

The problematic nature of estimating whale stock populations, and the need for change in the IWC's approach to uncertainty, was explained to the commissioners by COF scientist Sidney Holt at the 1964 meeting:

> We [the COF] would recommend that you avoid delaying action while you wait for better assessments. This has been done for the last decade or so. We have pointed out to the Scientific Committee that the best assessments are always made when the stock is thoroughly depleted. Our best estimates of what is happening to a stock of animals come when the stock has virtually disappeared and we have a nice long series of data showing its decline. If you wait until you have better estimates next year and better estimates the year after that you will only make decisions when it is much too late, certainly too late to act painlessly from the point of view of the whaling industry. Then, in any case, a new factor comes into consideration as it has with the blue whale, that you risk total extinction.[2]

The IWC's steady drift toward an entirely different paradigm for understanding and accounting for scientific uncertainty in policy making was made possible only by the then obvious dearth of the more profitable larger species the industry traditionally relied upon. The shrinking numbers of larger whales, reflected not only in the smaller catches but also by increasing catches of the smaller and less profitable sei whale, led to a further shrinking of the industry, not only in catches but also in the number of countries engaged in pelagic hunting in the Antarctic. This important change in circumstances underpinned a major transition in the attitudes of many commission members that began with the withdrawal of Great Britain and the Netherlands from pelagic whaling in the Antarctic—a transition that fundamentally altered the priorities of these governments and the many other IWC members that would withdraw from whaling over the next decade.

This period of change strongly indicates the extent to which such major shifts in attitude are dependent on similarly dramatic shifts in perceptions concerning related economic and political impacts, rather than simply on the strength and depth of the available scientific evidence. The important developments that occurred in the commission during the 1960s show quite clearly that evidence is unable to "speak for itself"—at least with a loud enough voice to be accepted by all concerned—and also that its potential meanings are only drawn from the interpretations of those with interests related to the evidence in question. Thus, individual interpretations of the scientific evidence presented to the IWC's members began to vary in accordance with the changes in their respective whaling interests, which by the mid-1960s were considerably less uniform than they had been a decade earlier.

Following the changes in the economic and political priorities of many, but certainly not all, in the IWC, the commission's management of what can be described as a "conservation interests versus industry interests" dialectic, which always has been the core issue in the commission's deliberations, was fundamentally altered during the decade following the Antarctic collapse. By the time of the moratorium's eventual adoption in 1982, it had become clear that the collapse of Antarctic whaling and the corresponding reduction in active whaling countries had created a vastly different political environment within the commission, one which was leading the IWC to charter an entirely new course for its approach to management.

In this chapter we will turn to the events of the 1960s that were behind this sea change (i.e., the heavy depletion of some species and the steady withdrawal of many countries from whaling), exploring how the change in attitudes gathered sufficient momentum by the end of the decade to cause a complete break with the commission's previously dominant perception of uncertainty. Former SC chairman Philip Hammond also has recognized the significance of this shift, attributing its origins to a fundamental change in priorities within the commission:

> Until [the IWC's adoption of the moratorium in 1982], the burden of proof had been on those seeking to limit whaling to demonstrate that a particular stock could not sustain the current level of harvest. Since then, it has been on those wishing to resume whaling to demonstrate that it is safe to do so. This is an important change of emphasis. It reflects the philosophy that cetaceans, and not the industry, should be given the benefit of the doubt because cetacean populations are more valuable in themselves than they are as a resource.[3]

The genesis of this important about-face resulted primarily from the demise of Antarctic whaling in the 1960s, a development that added considerable weight to the COT's warnings about delaying preventative action on the grounds of uncertainty. This chapter's discussion of the precursors to the first of many subsequent calls for a commercial whaling moratorium at the 1972 UN Conference of the Human Environment in Stockholm, therefore, will focus not only on the changing political and economic landscape within the IWC, but also on how these changes in perception created a situation within the commission that would lead the IWC to explicitly address scientific uncertainty and the many problems it presents for science-based approaches to policy making.

SCIENTIFIC UNCERTAINTY: AN EXCUSE FOR ALL SEASONS

The actions of the IWC's members at the 1964 meeting in Sandefjord clearly illustrate what this study is attempting to demonstrate: that the treatment of scientific advice by policy makers in the IWC (and in other wildlife and environment regimes) is determined almost entirely by how well it fits with individual priorities, rather than the extent (contrived, imagined, or otherwise) to which a piece of scientific research may or may not be said to accurately describe and explain "reality."

The meeting in Sandefjord was a moment of "truth" (to use the term loosely) for the IWC because the central issue at this meeting was the deadline for the commission to honor its pledge, made four years earlier, to set catch quotas in accordance with the advice and recommendations provided by the COT and SC. The opportunity to base its policies upon the "best scientific advice"—as per the ICRW and the commission's 1960 commitment to accept "science-based" management as the best strategy for avoiding the depletion of stocks and safeguarding the future of the industry—was within the IWC's grasp at this meeting, but was instead ignored in almost contemptuous fashion. This should have been no surprise, given the stalling that had occurred at the previous two meetings (in addition to the IWC's track record so far with conservation initiatives); the COF's recommendation's for drastic cuts in the Antarctic quota; and also the refusal of the Japanese, Soviet, and Dutch commissioners at the 1963 meeting to affirm the commission's earlier pledge to accept the recommendations of the COT.[4]

The outcome of the commission's 1964 deliberations on the quota for the 1964–65—in spite of the work done by the COT, COF, and the SC—represented little more than a return to the past attitudes and practices that had led Antarctic whaling into its current crisis. The beginning of a new era of cetacean management, as had been promised, was nowhere to be seen. The IWC's sixteenth meeting, therefore, ended in total failure.

In addition to the commission being unable to accept and implement any of the scientific advice it had promised to follow, IWC members also were unable to set any quota for the coming 1964–65 season. The remaining Antarctic nations were unwilling to reduce the catch below what they believed was needed to satisfy their whaling interests. According to the IWC's *Sixteenth Report*,

> the Commission were unable to agree upon a catch limit for the Antarctic pelagic expeditions for the 1964/65 season; those countries engaged in whaling could not see their way to accept such a drastic reduction of the 1963/64 catch limit as the scientific evidence indicated while the non-whaling countries were unable to vote for any limit substantially higher than warranted by this evidence.
>
> Since the limit of 10,000 blue whale units imposed by the Commission at their 1963 meeting related only to the 1963/64 season, their failure to agree in 1964 meant that there was no catch limit at all set by the Commission for the 1964/65 season.[5]

As noted in the closing section of chapter 3, Japan and the Soviet Union were the main opponents to the commission's implementation of policies based on the COF's report. These governments rejected the scientific committees' recommendations by arguing, firstly, that such major changes were financially untenable for the whaling industry and, secondly, that reductions of the magnitude recommended were not sufficiently justified by the evidence provided. The main SC recommendations in relation to the Antarctic stocks for the 1964 meeting were

(a) the use of individual quotas for each species instead of a single quota set in BWU;
(b) a reduction in Antarctic catches "to the lowest possible level below 4,000 fin whales and 5,000 sei whales" in order to avoid further depletion of these stocks; and
(c) the banning of all blue whale catches, including pygmy blue whales, in the Antarctic.[6]

The proposal to end the use of the BWU was as unpopular as ever and did not even rate mention in the Chairman's Report; such was the reluctance within the commission to even consider replacing a system that had shortcomings made increasingly obvious by the very nature of the COF's species-by-species analysis of current sustainable yield and potential maximum sustainable yield. Thus, even on this basic level (i.e., the issue of how catches should be counted and distributed), there was a huge gulf between what the scientific advice required and what the commission was prepared to accept and implement. The two main topics of debate at the 1964 meeting were the proposed elimination of Japanese and Soviet catches of pygmy blue whales and the reduction of the total Antarctic quota, which now consisted almost entirely of fin and sei whale catches.

The COF's conclusions in their 1964 report, completed just prior to the Sandefjord meeting, were given added weight by the close match between the committee's 1963 estimate for fin whale catches in the 1963–64 season and the actual catch taken in that season:

> In their Supplementary Report made to the Commission at its Fifteenth Meeting (Fourteenth Report, page 93), the committee predicted that "if whaling continued at the 1962/63 level of effort the catch in 1963/64 would only be 14,000 (fin) whales." In fact the catch was 13,853 fin whales though it should be pointed out that these were taken with 91 percent of the effort (catch days corrected for tonnage) expended in 1962/63. Thus the forecast made upon the basis of the previous calculations was correct to within 10 percent of the actual catch attained.[7]

With this strong correlation adding further credibility to their earlier calculations on the status of the Antarctic fin whales, the COF scientists reaffirmed their 1963 estimate of an Antarctic fin whale population of about 40,000 and reduced this figure to between 35,000 and 36,000 for the 1964–65 season after adjustment for the recruitment of new adults (born in earlier seasons when the number of whales was higher) to the population. Working with the assumption that the sustainable yield rate of fin whales (r–M) was 0.12, the COF estimated "the sustainable yield from the present stock [i.e., 35–36,000 animals] to be between 4,000 and 5,000 whales. Catches much in excess of this number will cause continued rapid depletion of the stocks. Catches below this level will allow rebuilding of the stocks to increase future productions."[8] This meant that a maximum of only 2,000–2,500 BWU could be provided by fin whales, which had become the mainstay of Antarctic whaling following the depletion of the blue whale stocks.

In order to offset the growing scarcity of fin whales, the Antarctic nations had begun taking increasingly bigger catches of the smaller sei whale. This increase, however, did not match the drop in the number of other species caught (blue, fin, humpback) or even the drop in only the fin whale catch, which fell by 4,783 in the 1963–64 season; the number of sei whales taken increased by less than 3,000.[9] In the COF's view, and also that of the SC, the inability of some of the whaling fleets to maintain their quotas by switching to sei whales (only Japan and the Soviet Union met their quotas in 1963–64)—as they had done in the 1950s with fin whales offsetting the decrease in blue whales—supported an earlier contention made in the COT's "Supplementary Report":

> The sei whale stock is not big enough to sustain "catches that will support an industry comparable to that which has been operating in the Antarctic," since it may be presumed that if sei whales were abundant the expeditions would take them in at least equivalent numbers when the other species are scarce.[10]

The main data sources used by scientists to make stock and sustainable yield estimates in the postwar period were primarily CPUE figures in addition to biological information (based on estimated pregnancy/recruitment and natural mortality rates, age determination methods, and so forth); preexploitation population estimates based on catch history; whale marking data; and also the results of a much earlier Antarctic sighting survey conducted between 1933 and 1939. So although the COT and COF were able to improve both the collation and analysis of these data, and then provide more detailed and better supported estimates for the Antarctic blue, humpback, and fin stocks than ever before, there still remained considerable uncertainty concerning even the most fundamental information upon which their conclusions relied (namely, how many whales are there now? How many were there before? What are the appropriate rates for recruitment and mortality?).

In the case of the blue and humpback whales, which had been so heavily hunted in the past, a large body of data from previous seasons was available for the COT scientists to track the decline in CPUE, which became the mainstay of their conclusions. The importance of any doubts over the accuracy of the data mentioned above—at least concerning whether quota reductions were required—was largely negated by the inability of the fleets to find these animals in any significant numbers despite the use of faster and more efficient catcher boats.

Faced with such a situation, by the mid-1960s it was difficult to argue against these species' decline. This suggests that the only empirical evidence of unsustainable hunting that could be regarded by all in the IWC as reliable was the apparent scarcity of a species—something that only became obvious after serious depletion of the stock already had occurred.

This approach to management gave the benefit of the doubt in estimates to the industry and was, therefore, clearly an advantageous way of perceiving scientific uncertainty for the active whaling nations vis-à-vis their criterion I priority (i.e., catching as many whales as possible). As a result, this style of thinking remained prevalent among those IWC members who mattered the most—the remaining Antarctic nations—when adopting management initiatives. This created a significant obstacle to the remainder of the commission's attempts to avoid past mistakes by adopting a more conservation-oriented interpretation of uncertainty.

The situation facing the Antarctic sei whale stocks thus resulted from the increasingly desperate circumstances facing the whaling nations in their quest to catch what they believed was "enough" whales. Sei whales were coming under increasing pressure and the major concern among scientists was that these stocks would also be depleted. The COF had very little data on these animals, since they had not been hunted in any significant numbers until the late 1950s, and this made it impossible to develop the same detailed analysis of their condition that had been presented for the blue, humpback, and fin whales. The COT's "Supplementary Report" to its "Final Report," considered by the commission in 1963, noted,

> The catch of sei whales increased from 4,716 in 1961/62 to 5,482 in 1962/63. At the moment the committee has no biological data on sei whales nor any way of measuring the effective effort applied to sei whale catching so that there is no way of determining total population size or sustainable yields. However, it seems that this population is unlikely to be able to sustain continued catches that will support an industry comparable to that which has been operating in the Antarctic.[11]

But by the following year, with sei whale catches continuing to increase markedly, the COF felt compelled to provide some estimates on the condition of this species. The committee remained hamstrung by a lack of data, both biological and catch-related, and acknowledged this very clearly in its 1964 report while again stressing the need for the BWU system to be replaced with species by species management:

> The catch of sei whales increased this season by nearly 3,000 whales, from 5,842 in 1962/63 to 8,256 in 1963/64. . . .
>
> The need for consideration of the sustainable yield for this species will be further accentuated if, as the Committee recommends, the blue whale unit is abolished so that separate quotas for each species have to be considered. The small amount of available data for this species makes possible only an approximate assessment of the situation. A lower limit of the present sei whale population can be obtained on the assumption that under the conditions prevailing last season when quotas were obtained with difficulty, if at all, the ratio of sei to fin whales in the catches represents the ratio of their populations. The estimate obtained is 20,000 sei whales. This estimate could be subject to correction for sei whales occupying areas in the Antarctic where whaling did not take place, and an alternative approach can be obtained from the sighting survey of 1933–39. At this time the proportion of sei whales seen was very low in relation to the other species, although there were difficulties in observing and recognizing this species. The total fin whale population at that time is estimated at about 125–150,000, so that 50,000 would seem a high estimate of the corresponding sei whale population. It is possible that some increase and some southward movement has occurred in this species as a result of relaxation of competition from other baleen whales, but on the other hand 26,000 sei whales have been taken in the last five years. An extreme upper limit of the present sei whale population would therefore seem to be 70,000. Taking, in the absence of specific data, the best estimate of (r–M) for fin whales (0.12) as applying to sei whales, the present sustainable yield for this species is between 2,400 and 8,400 whales, with a probability of being nearer the lower limit. Since the actual catch in 1963/64 (8,256) almost exactly equaled the upper limit, it is likely that the sei whale population is already being reduced, and that any further increase in catches will inevitably deplete it at a rapid rate.[12]

At the 1964 meeting, Japanese and Soviet opposition to reducing the quota to 4,000 BWU and protecting pygmy blue whales—proposals that had majority support within the commission—was based firstly on the need to catch more whales and secondly on doubts concerning the necessity of such measures. Given the lack of data on sei whales, the COF scientists could only guess at the numbers and potential yield, and their estimates probably were influenced by the history of species-by-species depletion that so far had characterized the whaling industry's management record.

What they, and now many others in the commission, were pushing for was a more precautionary approach (i.e., in favor of the resource rather than the industry) to the interpretation of the many uncertainties that unavoidably accompanied assessments.

However, such a fundamental change in perceptions was only attractive to those who believed it was a change they could afford to embrace; the Japanese and Russians clearly felt the price was too high and so questioned the COF's conclusions. In a criticism reminiscent of the Dutch posturing during the fin whale debate, Soviet scientist Tverianovitch told the SC in 1964 that "estimates of a sustainable yield especially for sei whales are not convincing enough, because the condition of stocks of this species appears to be more than satisfactory."[13]

After considering the SC's recommendations, based on the COF's findings, that the Antarctic quota should be based on catches of *less than* 4,000 fin whales and 5,000 sei whales (i.e., less than approximately 2,830 BWU) for stocks to reach their optimum level for MSY,[14] the Technical Committee recommended to the commission quotas of 4,000 for the 1964–65 season and subsequent reductions to 3,000 and 2,000 in the following two seasons.[15] Japanese commissioner I. Fujita responded, "Japan cannot discuss this problem [i.e., the Antarctic quota] only in terms of the conservation of whale stocks."[16]

Japan's position made clear the huge gap between the remaining Antarctic whaling countries and the rest of the commission concerning the long-term benefits of greater conservation. This division was further reflected by the Soviet commissioner, Captain A. N. Solyanik, who echoed Tverianovitch's opinion by saying that "the conditions in the Antarctic area are not as bad as has been presented to us."[17] Solyanik also noted that both the Japanese and Soviet fleets had reached their respective quotas in the previous season (only, it can be argued, because the 1963–64 Antarctic quota had been reduced by 6,000 units to 10,000 BWU).

Indeed, the Japanese and Soviet view of the situation in 1964 was based on their requiring the kind of evidence that was only likely to be provided after the fin and sei whale stocks had been depleted to very low levels, as had occurred with the blue whale. This style of thinking also was made evident in the SC, where the Japanese and Soviet scientists had been alone in contending that a drastic reduction in the Antarctic quota was not based on "sufficiently reliable data."[18] The problem then, in terms of the commission being able to act upon the scientific advice it had asked for and received, was the reduced but still significant influence that industry priorities maintained over the interpretation of the evidence generated by scientific research.

With the decline of the whaling industry had come a corresponding decline in the number of commissioners willing either to dispute the validity of the scientific advice at hand or risk further overhunting. This trend was made conspicuous not only by the newly found commitment of many in the IWC to following the recommendations of its scientists, but also by the absence of the Netherlands in disputes over the reliability and interpretation of scientific data in both the plenary and the SC.

As also was noted in chapter 3, the Netherlands already was involved in negotiations with a Japanese company over the sale of its only factory ship, the *William Barendsz*, prior to the 1964 meeting. The sale went ahead a short time after the meeting, following Japan and the Soviet Union's success in resisting the implementation of a drastically lower quota for the coming season. At the 1964 meeting, therefore, the Netherlands' interest in the Antarctic quota appears to have been more concerned with what would satisfy the Japanese rather than any expectations for its own catch.

What transpired in the plenary sessions effectively was a struggle between, on the one hand, the governments that no longer (or had never) maintained Antarctic pelagic whaling interests and were, therefore, pushing for the adoption of the 4,000 BWU limit as per the 1960 commitment and, on the other hand, the governments that were planning to continue with Antarctic pelagic whaling. Since the United Kingdom already had sold its last expedition to Japan and now only hunted from a land station in South Georgia, and the Netherlands also was about to sell its only expedition to Japan, the latter category of active pelagic whaling countries in the plenary actually consisted of only Japan, the Soviet Union, and Norway.

For its part, Norway supported a compromise of 6,000 BWU (for the 1964–65 season only), stating that while Norway would accept the 4,000-unit limit, the government believed there was no point in doing so as this proposal clearly would be objected to by the Japanese and Soviet governments, since they were demanding a minimum of 8,500 BWU.[19] The various proposals for the Antarctic quota made in the plenary and the voting that followed were summarized in the meeting's Chairman's Report, which illustrates how the proceedings were ultimately influenced by the immediate needs of the remaining industries rather than by the scientific advice at hand:

Bearing in mind the scientific advice, the Commissioner for the United States seconded by the Commissioner for Australia proposed that Paragraph 8(a) of the schedule should be amended as follows: "The number of baleen whales taken during the open season caught in waters south of 40[dg] South Latitude by whale catchers attached to factory ships under the jurisdiction of the Contracting Governments shall not exceed 4,000 blue whale units in 1964/1965, 3,000 blue whale units in 1965/1966 and 2,000 blue whale units in 1966/1967."

To this substantive motion was added the following amendment proposed by the Commissioner for Japan and seconded by the Commissioner for the Union of the Soviet Socialist Republics: "The number of baleen whales taken during the open season caught in waters south of 40[dg] South Latitude by whale catchers attached to factory ships under the jurisdiction of the Contracting Governments shall not exceed 8,500 blue whale units in 1965/1965."

In its turn this amendment was subjected to the following amendments proposed by the Commissioner for Norway and seconded by the Commissioner for Iceland: "The number of baleen whales taken during the open season caught in waters south of 40[dg] South Latitude by whale catchers attached to factory ships under the jurisdiction of the Contracting Governments shall not exceed 6,000 blue whale units in 1965/1965."

On being put to the vote the second amendment was lost by 12 votes to 1, with 1 abstention, the first amendment was lost by 11 votes to 3, and the substantive resolution was lost by 10 votes to 4, the 4 countries voting against it being those at present engaged in Antarctic whaling [i.e., Japan, the Soviet Union, the Netherlands, and Norway]. Somewhat later in the meeting the Commissioner for the Netherlands seconded by the Commissioner for Norway proposed a further amendment, that: "The number of baleen whales taken during the open season caught in waters south of 40[dg] South Latitude by whale catchers attached to factory ships under the jurisdiction of the Contracting Governments shall not exceed 8,000 blue whale units in any one season provided that no lower catch limit is established."

This proposal was also lost 10 votes to 3 with 1 abstention.[20]

By the end of the meeting, no proposal for the setting of the 1964–65 season's Antarctic quota had gained the required three-quarters majority, leaving the commission with no quota at all as the previous quota of 10,000 BWU imposed at the 1963 meeting related only to the 1963–64 season. This stalemate over the Antarctic quota was reminiscent of the deadlock that had occurred several years before over the national quotas scheme, with the main obstacle to agreement again being the threat a significant quota reduction would pose to the industry's financial interests and the political well-being of the governments concerned. And like the national quotas issue, agreement on some form of catch limit finally amounted to the Antarctic whaling countries giving themselves what they had demanded, with the remainder of the IWC reduced to looking on from the sidelines.

In the absence of a quota set by the commission, the commissioners representing the four remaining Antarctic whaling nations (with the Netherlands in name only) simply agreed between themselves at the end of the 1964 meeting to recommend to their governments a voluntary catch limit of 8,000 BWU for the 1964–65 season (i.e., essentially the same conditions as in the final amendment proposal made earlier by Norway). In its concluding summary of the commission's voting on the various proposals for the 1964–65 Antarctic quota, the Chairman's Report observed,

> The resolution proposed by the Commissioner for the United States was compatible with the scientific advice which the Commission at their meeting in 1960 had undertaken to implement by July, 1964. The case of those who voted against this resolution, however, was based on Article V of the Convention, which states that regulations with respect of the conservation and utilization of whale resources "shall take into consideration the interests of the consumers of whale products and the whaling industry" (Paragraph 2(d)). Some of the opponents contended that the condition of their whaling industries could not support the catch restrictions recommended.[21]

The COF and SC's recommendation for the protection of the pygmy blue whale stock, still being hunted by the Japanese and Soviets, fared no better than the attempt to reduce the total quota, and again the main opponents to this measure were the two governments that stood to lose the most from its adoption. At the 1963 meeting, complete protection for all Antarctic blue whale stocks (including the pygmy) was opposed firstly by the Japanese and Soviets on the grounds of a lack of available data on the pygmy blue whale and then by the British commissioner, who argued that complete protection would risk objections from Japan and the Soviet Union and result in all Antarctic blue whales being at risk. Interestingly, the threat of objections was no longer an issue for the United Kingdom in 1964, after it sold its last remaining factory ship to Japan; at that meeting the British commissioner proposed protecting the pygmy blue whale in addition to the larger blue whale, a significant change in position from the previous year.

But although the British proposal to ban all catches of blue whales in the Antarctic received the required majority (only Japan and the Soviet Union voted against), the measure could not be implemented due to Japan and the Soviet Union, followed by all the Antarctic nations plus the United Kingdom, lodging objections. The Japanese and Soviet commissioners had supported decreasing the area in which the pygmy blue whales could be caught, but refused to stop catching altogether. As so far had been the case in the IWC's treatment of uncertainty issues concerning stock numbers and sustainable catch, those with interests at stake—again the Japanese and Soviet delegations—took the continuing absence of data on the pygmy blue whale stock to mean that there was no scientific reason to stop hunting. And so the debate over the pygmy blue whale catch resembled the Antarctic quota debate in that it again illustrated the fundamentally different perceptions of uncertainty and the risks it involved.

But in the case of the Antarctic quota, the debate only concerned reducing the catch to the estimated sustainable yields or below, and did not involve proposals for the complete protection of fin and sei whales. The Japanese and Soviet commissioners were disputing the size of the reduction required rather than the need for some reduction, and they resisted proposals lower than the 8,500 BWU they had demanded primarily on the grounds of what their fleets needed to catch. And although the Soviet position included some references to the incomplete nature of the data used by the COF—particularly in regard to the sei whale stocks—the large amount of evidence compiled by the committee to support its contention that fin whales were under threat was difficult to dismiss entirely. This made it more difficult to argue against the COF's advice on uncertainty grounds alone, especially in light of declining catches.

However, the SC's 1964 recommendation for the pygmy blue stock was for complete protection rather than a reduction. This made the issue even more contentious given the considerable uncertainty over the stock's sustainable yield, as with the sei whale stocks, and also the COF's earlier agreement with the Japanese scientists that "the stock of pygmy blue whales could stand the annual catch of 400 whales for three seasons."[22] The biggest problem with continuing catches, in the Scientific Committee's majority view, was the risk such catches would present to the Antarctic's remaining blue whale population. The majority of scientists including the COF, with the Japanese and Russian scientists dissenting, believed that further pygmy blue whale catches would lead to accidental catches of "large blue whales" because "pygmy blue whales cannot be identified in the water."[23] The SC report explained,

> The number of large blue whales which would be included in a total catch of 400 blue whales in the area open at present can roughly be predicted from the ratios of pygmies to large blue whales in the catches in that area in recent seasons. This ratio varies from month to month and square to square, and it would be expected to vary from year to year, but for the Japanese catches in 1963/64—the only data available to the Committee—the ratio was 31:7. The expected unavoidable catch next season of large blue whales in a total of 400 would therefore be about 70, and this is certainly much greater than the already extremely low sustainable yield of the Antarctic stock....
>
> The Commission must therefore balance the risk—admittedly incalculable but nevertheless real—of a permanent loss of a potential 6,000 large blue whales a year [i.e., after the stocks recovered to the point where estimated MSY could be achieved] against an annual catch of a few hundred small pygmies for a very limited period.
>
> With these considerations in mind this Committee concludes that, for economic reasons alone, there is a very strong case for the immediate and complete cessation of catching blue whales, including pygmies, and recommends that the commission act accordingly.
>
> Furthermore, whatever the economic considerations, the Committee's members, as scientists, believe there is no justification for increasing the serious risk of extinction of the main stock of the largest living animal. In this they are in company with other scientists and organizations, several of which [including the UN Food and Agriculture Organization] have strongly expressed this view publicly and in letters to the Commission.[24]

But Japan and Russia simply were not prepared to accept the magnitude of the risk to blue whales posed by further pygmy blue whale catches. The rationale for rejecting the warning, in terms of how uncertainty was perceived, was again in line with the whaling countries' refusal to accept that greater certainty over the risks and costs involved could only be had at the expense of further risk to the stocks and, ultimately, to the industry itself. Japanese scientist Omura's response to the committee's view interpreted the associated uncertainties by giving the benefit of the doubt to the industry and its desire to catch as many whales as possible:

> Judging from the present status of pygmy blue whales in the Antarctic, there is no need yet for taking measures for their total protection. While it cannot be denied that there is a slight possibility of taking ordinary blue whales in the permitted area, the statistics for the past five seasons *prove* that the number of ordinary blue whales caught in the said area was *very small*.[25]

In addition to the question of why Omura was able to base his statement on catch data from "the past five seasons" while the SC had available to it only the 1963–64 catch figures, Omura's opinion in the SC is made even more curious by the reasoning he uses while attempting to justify further catches. Even if we leave aside his vague use of the phrase "a slight possibility" in relation to the chances of previous large blue whale catches being repeated, Omura's proposition that he can *prove* something is "very small" betrays a lack of logic that was most likely forced by the political imperatives of being a scientist in the employ of a government determined to continue catching. The meaning of "very small" clearly can only be relative, and Omura's use of the phrase here seems a rather glib attempt at downplaying the uncertainties that were necessarily involved in his own argument, providing a good example of the selective way in which the whaling nations either ignored or highlighted uncertainty depending on how its existence affected their interests.

Fujita, the Japanese commissioner, provided another such example in the plenary by arguing that a catch of 1,200 pygmy blue whales over three seasons would not cause "any harmful effect upon the stock." This assertion was then contradicted by FAO scientist Holt, who explained that although the COF concluded the stock "could stand" such a catch, the committee did not state that "this could be done without causing any harmful effect." Fujita also cited Omura's opinion on the issue of incidental large blue whale catches but downgraded his assessment of "slight possibility" to "no risk." This assertion again was dismissed by Holt on the basis of the COF and SC's opinion.[26]

Thus, when the time came for all in the IWC to acquiesce in the setting of policies based on the findings of the COT and the SC, a move that would have meant a fundamental change in each member's perception of uncertainty and how it should be dealt with (i.e., precaution in favor of the stocks rather than the industry), the governments that stood to lose the most by doing so quickly retreated from their earlier pledge to follow the advice provided. That Japan and the Soviet Union were the only two governments with a long-term commitment to continuing pelagic whaling in the Antarctic, and were also the two main obstacles to the implementation of the advice provided by the COT, should not be seen only as mere coincidence. Indeed, their refusals to accept the majority view that short-term profits could only be pursued at the expense of far greater long-term benefits (i.e., larger yields from larger stocks) clearly indicate the extent to which the treatment of science and the advice it provides is ultimately determined by criterion I and criterion II priorities. This proposition is further supported by the change in attitude among other members that occurred only after the disposal or marginalization of their whaling interests.

AND THEN THERE WERE TWO . . . THE END OF AN ERA

The remainder of the 1960s saw the continued decline of the Antarctic whaling industry with the Soviets, Japanese, and particularly the Norwegians obtaining smaller and smaller returns as the fin and sei whale stocks dwindled.[27] The remaining Antarctic nations numbered only three in the 1964–65 season following the sale of the Dutch factory ship, the *Willem Barendsz*, to Japan in October 1964. The sale increased Japan's allowed catch, under the now revised Antarctic quota allocation, to 52 percent, or 4,160 BWU. The total catch for the season was only 6,986 BWU with more than half of the BWU coming from sei whales (19,874 sei whales against only 7,308 fin whales)[28]—considerably less than the 8,000 BWU the Antarctic nations had awarded themselves following the deadlock at the 1964 meeting. And although both Japan and the Soviet Union (20 percent) ended the season just short of their already reduced quotas, the big loser was Norway, whose two remaining fleets managed only slightly more than half of their 2,240 BWU quota, which had remained at 28 percent.[29]

The following year's meeting in London, in June 1965, was preceded by a special meeting in May, convened in response to the 1964 meeting's failure to set any Antarctic quota "in order to obtain some agreed total quota which would allow the restoration of the Antarctic whale resources."[30] The upshot of this special meeting was that a Committee of Six was formed, consisting of the three Antarctic pelagic countries plus the United States,

Canada, and France. This committee eventually agreed upon a quota of 4,500 BWU for the 1965–66 season only, after debating various quota proposals for the 1965–66, 1966–67, and 1967–68 seasons.

Not surprisingly, the United States proposed the biggest reductions in the total quota (3,000, 2,000, 2,000 BWU for each season), while Japan supported the smallest reduction (4,500, 4,000, 3,500 BWU for each season). Norway, again taking the middle ground, argued for 4,000, 3,000, and 2,000 for each season. However, both Norway and Japan attached conditions to their proposals. Norway insisted on catch distribution being determined by "a national quota system" (implying Norway's willingness to accept Soviet demands that the earlier scheme be renegotiated). Japan required that "the national quota system" (i.e., the current system) be "in operation" and also demanded that the long awaited International Observer Scheme (IOS),[31] which the five Antarctic nations had agreed to in principle in 1963 but were still unable to implement due to additional demands made by the Soviets, finally be implemented.[32]

Attempts at reaching agreement at the special meeting were further complicated, as had been the case at the 1964 meeting, by uncertainties concerning sei whale abundance estimates and the sustainable yield of these stocks. Japan and the Soviet Union remained unwilling to accept the COF's advice concerning a drastic and ongoing reduction in the Antarctic quota that would reduce the total catch to below the committee's sustainable yield estimates.

At the 1964 meeting, the IWC had requested that the Fisheries Division of the FAO continue the work of the COF, which by this time had completed its assessment of Antarctic blue, fin, and humpback stocks and provided advice on what it believed to be appropriate measures for achieving MSY from these stocks in the shortest possible time. The FAO, which supported the COF's conclusions, accepted the commission's request but only on the condition that the IWC would refrain from pursuing the industry's short-term financial needs at the expense of the recovery of stocks and the industry's long-term viability. Speaking on behalf of the FAO director general at the IWC's 1964 meeting, Holt informed the commission,

> I have been asked to say that naturally F.A.O. cannot undertake to collaborate in scientific work the results of which are to be used merely to organize the more efficient destruction of the resource. Thereby F.A.O. would itself become associated with such a policy, and we cannot do that. While, however, there remains any hope that international agreement to

work towards a high sustained yield can be reached and put into effect, F.A.O. will do everything possible to help to ensure that the best possible scientific advice is available for this purpose.[33]

The FAO by this stage clearly had become very concerned with the plight of the Antarctic stocks, and in particular with the IWC membership's inability to honor its 1960 pledge to follow the COT's advice. As the 1965 special meeting demonstrated, Japan and the Soviet Union still were unwilling to cut catches to the levels recommended by the COF, choosing instead to play the uncertainty card over sei whale numbers and catch levels in order to defend their position and push for a higher quota. As the IWC report of the special meeting states, industry interests remained largely unaffected by the advice on offer—a situation that clearly reflected Japan and the Soviet Union's ongoing preference for giving the benefit of the doubt to their industries rather than the resource:

> The International Whaling Commission recommends to the Governments party to the International Whaling Convention a total quota of 4,500 B.W.U. for the Antarctic season of 1965/66. The Commission recognizes that this quota will not in the 1965/66 season reduce the catch below the sustainable yield as determined on the basis of the scientific evidence available, *which is incomplete at least in regard to sei whales*. This quota is agreed as a transitional limit to assist the pelagic whaling industries to adjust to the reductions required to begin rebuilding the whale herds. All members of the commission agree that they will recommend to their governments that they support further reductions for the 1966/67 and 1966/68 seasons that will assure that the quota for the 1967/68 season will be less than the combined sustainable yields of the fin and sei stocks as determined on *the basis of more precise scientific evidence*.[34]

Thus it was with the expectation that the IWC would, sooner rather than later, bring its catch quota in line with the available scientific advice that the FAO agreed to provide scientific support to the SC. In 1965, during the IWC special meeting, the FAO submitted a report emphasizing, as had the reports of the COF and SC, the need for quotas to remain at levels below current estimates of sustainable yield. The report also addressed the prevailing uncertainty over the sei whale catch, while also warning against the continued depletion of sei whale stocks as the Antarctic nations sought to fill the void left by fin whales. In the FAO's view,

The only policies which can be considered as meeting the objective of recovery and rational exploitation of the resource are those which lead within a specified period of time to catches of each species not exceeding sustainable yields and the stocks increasing towards the levels at least as high as those giving maximum sustainable yields. . . .

Firstly it may be said that the only policy which would imply an immediate step towards conservation in the 1965/66 season is one that would ensure that: (a) the fin whale catch did not exceed the present sustainable yield of about 4,000 whales = 2,000 B.W.U., (b) the sei whale catch did not exceed its present sustainable yield, and (c) capture of blue whales was prohibited.

With respect to element (b) there will be, in the coming year, a recruitment which is greater than the present stock will produce, as it is coming from the larger stock of earlier years. Now that the sei whale stock has been rather drastically reduced, the recruitment of future years will be correspondingly reduced. Thus the catch of sei whales in 1965/66, which would leave the stock of legal sized whales virtually unchanged, is something rather more than the value of about 3,000–4,000 (= 500-650 B.W.U.), which is estimated as the sustainable yield in the steady state condition of a stock of the present size (which is, it seems, near the optimum level); how much more could only be determined by a detailed analysis of all available data for sei whale as recommended by the Committee of Four.

We can say then that only if the total catch in 1965/66 were less than 2,500 BWU would it be ensured that both fin and sei whale stocks were not further depleted.[35]

When the commission met in June, with both the FAO report in hand plus an SC report that "regretted" the adoption of a 4,500 BWU quota and unequivocally supported the FAO's advice,[36] it was becoming increasingly difficult for the Antarctic nations to ignore demands for IWC policy to follow the available scientific advice. This was due to both the growing political pressure for catch reductions from both within and outside the commission, and also the straightforward and relatively unambiguous language the scientific advice was now being framed in. Indeed, the Soviet and Japanese dissent within the SC—which only one year earlier had prevented complete protection for all blue whales, including pygmy blue whales, in the waters south of 40 degrees south latitude—seemed to have disappeared by the committee's 1965 meeting. Instead, there was a unanimous call for the IWC to "extend the area in which the taking blue whales is prohibited to include the whole of the southern hemisphere—as is the case for humpbacks."[37]

However, initiating further protection for blue and humpback whales, which by 1968 included all of the Southern Hemisphere, the North Atlantic, and the Pacific,[38] was the closest the IWC came to implementing the advice of its scientists in the Antarctic at the 1965 meeting. It would be 1967 when the commission voted to accept an Antarctic quota for the 1967–68 season of 3,200 BWU, a further reduction from the 3,500 adopted at the 1966 meeting.[39] By the 1967 meeting, the SC, on the basis of a recent FAO Assessment Group report (which included 1966–67 season catches), had revised its earlier sustainable yield recommendation for fin and sei whales in the Antarctic from 2,500 to between 3,100 and 3,600 BWU (4,800 fin whales and 4,400–7,000 sei whales).[40]

But while the Antarctic quota now finally matched what the SC had agreed to be a sustainable level of hunting, the move still ignored the committee's calls for catches to be set much lower than the sustainable yield level in order to allow stocks to recover. Furthermore, the reduction was, in any case, probably agreed to by the Japanese and Soviets (who both had wanted 3,300 BWU[41]) mostly because of the steadily declining Antarctic catches. Indeed, even though Norway, Japan, and the Soviet Union each managed to reach their respective quotas in the 1966–67 season (the last time the total Antarctic quota for a season would be reached), catch per catcher days for the season had dropped sharply to 0.30 BWU from 0.40 only two seasons before in 1964–65.[42]

The 1968–69 season quota remained at 3,200 BWU, but only 2,469 BWU were taken under a new system of national quotas, renegotiated again in 1968, which reduced each nation's share.[43] All of the three Antarctic nations failed to make their respective quotas, with Norway recording the lowest catch with only 292 of its now 731 BWU quota.[44] Consequently, this was the last time Norway sent a fleet to the Antarctic, effectively marking the end of Norwegian involvement in an industry the Norwegians had pioneered. With whale oil prices continuing to fluctuate wildly, the Norwegians could no longer justify the huge cost of sending a fleet to the Antarctic given that there were very few whales left to hunt.[45]

In addition to the declining catch per catcher-day figures, the ratio of sei to fin whales taken in 1967–68 (4.8:1) stood at almost half of what it had been in 1965–66 (7.6:1), a strong indication that sei whale stocks indeed had been hunted beyond the species' sustainable yield. By the 1969–70 season, the ratio stood at only 1.9 sei whales for every fin taken,[46] a further decline that soon would result in the Japanese and Soviet fleets turning to catches of Bryde's whale and the much smaller (and less profitable) minke whale. In a final throw of the dice, Norway experimented with a combined factory/catcher ship in the 1969–70 season, after having sent no ships at all in the previous season, in an effort to reduce costs to a level where some profit could be realized.[47] Norway's attempt at salvaging something from the ruins of Antarctic whaling, however, failed. The Norwegians left the Antarctic for the last time after managing only a meager total of 6 BWU (4 fin and 22 sei whales).[48]

During the late 1960s and early 1970s, the SC's deliberations on sustainable yields and suitable quotas largely focused on disagreements concerning recruitment and mortality rates for fin and sei whales and the effect these uncertainties had upon the committee's sustainable yield estimates and quota recommendations. Although the makeup of the SC had by now changed to include many more population dynamics specialists than before, and also was supported (until 1970) by FAO research, fundamental uncertainties concerning the biological data needed for abundance and sustainable yield estimates remained. This left the data open to a variety of interpretations concerning both what the available information should be taken to mean and also the conclusions that could be drawn from it.

In 1969, the SC, after adopting a new interpretation of ear plug growth layers the previous year, informed the commission that it had found errors in the COT's earlier fin whale estimates; the SC concluded that the COT had underestimated the total fin whale population but also had overestimated the net recruitment rate.[49] But these errors appeared largely to cancel each other out in terms of the COT's final conclusions concerning sustainable catch limits and the need for catches to be further reduced. The revisions were therefore of little consequence, due to the significant differences of opinion that were now developing in the SC over fin whale abundance and yield. Such was the magnitude of these differences, primarily between the Japanese scientists and the remainder of the committee, that the commission decided at its 1969 meeting to further reduce the 3,200 BWU Antarctic quota:

> The Scientific Committee had agreed on an estimate of 5,000 whales (833 blue whale units) for the 1969/70 sustainable yield of sei whales in the Antarctic but was unable to reach agreement on the yield of fin whales for which the estimates ranged from 1,000 to 5,600 whales. The difference arose in methods and assumptions and the Scientific Committee proposed that

it should meet early in 1970 to discuss and review methods and assumptions of fin whale stock assessment when these questions would be studies in greater detail and more precise estimates obtained. The Commission accepted the Technical Committee's recommendation that pending the more precise estimates to be made at the meeting early in 1970 the Antarctic catch limit in 1969/70 should be 2,700 blue whale units.[50]

Former COF members Allen and Chapman both had used different methods to arrive at similar conclusions over sustainable yield and population numbers for fin whales. Allen, using data from the North Pacific stocks, had developed a method that estimated "the gross recruitment from the age composition of the catch and in turn uses these estimates together with the catch and effort data for the seasons 1945/46 to 1964/65 to calculate stock size and yields."[51] On the basis of his analysis, Allen believed the current fin population to be 60,000, with a 1969–70 sustainable yield of 1,000–2,800 animals. Chapman, meanwhile, had employed a method that depended on decline in CPUE and a rate of net recruitment based on the revised age-reading method using ear plugs; he concluded the fin whale population to be 56,000, with a sustainable yield of 2,900.[52]

However, the Japanese scientists, led by Doi, came up with markedly different results "using methods and parameters such as age composition for the period 1931/32 to 1967/68 based on new age-length keys, age at recruitment and sexual maturity, marking and sighting."[53] In the opinion of the Japanese scientists, the fin whale population in 1969 was somewhere between 45,000 and 95,000 animals, affording a sustainable yield for the 1969–70 season of between 3,400 and 5,600.[54]

Most of the SC members and the FAO scientists believed the best estimate to be between 1,000 and 5,600, but the Japanese scientists rejected these figures, claiming the true number to be between 4,900 and 5,600 while also conceding that it was "not possible to give a single estimate within that range."[55] With the catch of fin whales now increasing in relation to the catch of sei whales,[56] the status and yield of the fin stocks again were of great importance to the future of Japanese and Soviet whaling in the Antarctic. Consequently, it is not surprising that the Japanese scientists continued to insist on higher abundance and sustainable yield estimates at the 1970 special assessment meeting (which had been convened to attempt resolution of the disagreement that had arisen over the previous year's estimates). At this 1970 meeting, unanimous agreement within the SC over fin whale numbers remained elusive due to the Japanese scientists' insistence that the sustainable yield was considerably higher than the estimate accepted by the rest of the committee:

> Much discussion was given to narrowing the differences of the estimates provided by the different scientists but the committee was unsuccessful in reaching a single estimate for the sustainable yield in 1970/71 because of a lack of good direct evidence. All members *except Japan* agreed that the recent level of fin whale catch in the Antarctic (2,700 average over the last five seasons) appears fairly close to the present sustainable yield. Japanese scientists *believe* the best estimate for 1970/71 is 3,250 to 4,350.[57]

The ongoing uncertainty and disagreement within the SC resulted in the Antarctic quota remaining at 2,700 BWU until the 1971–72 season, when it was reduced to 2,300 BWU. Why the Soviets and in particular Japan were prepared to accept the reduction from 3,200 BWU without lodging further objections on the basis of uncertainty in the data is an important question, but it is also one that can be answered relatively simply. Although the quota indeed was now significantly smaller, this reduction in the total allowable catch had no effect on the Japanese and Soviet catches because the Norwegians were no longer taking any of their quota. Therefore the reduction to 2,700 BWU represented no real reduction at all, since the Soviets and Japanese were still entitled to take the same number of whales allowed under the 1968 renegotiation of the national quotas agreement (i.e., Japan 1493 BWU, the Soviet Union 976 BWU).

So while the latter years of the 1960s finally saw the IWC bring its catch limits into line with the scientific advice for the first time in the commission's history, this milestone should not simply be attributed to the improvement in assessment methods or any resulting increase in influence on the part of the IWC scientists. The real reason behind the precipitous drop in the Antarctic quota was the equally precipitous drop in the abundance of the whale species Antarctic whaling relied upon. There was no new-found willingness on the part of the whaling nations to give the benefit of the doubt in uncertainty issues to the stocks instead of their industries.

The SC, by the late 1960s, was certainly providing much more detailed advice on what the Antarctic catch limits should be. But at the same time the Japanese, and to a lesser degree the Soviets, were still undermining scientific advice with arguments about the uncertainty of the data in the same way that the five Antarctic nations (with the Dutch at the forefront) had used uncertainty in the 1950s and early 1960s to justify resistance to quota reductions. The only real difference in how uncertainty was being perceived by the mid- to late 1960s stemmed not from the apparent willingness of governments to accept a proresource precautionary approach as a management principle, but from the absence of reasons to reject it.

Another example of the extent to which the interpretation of scientific uncertainty still was being determined by industry interests (or the lack thereof) was the continued use of the BWU system. The replacement of the BWU system with species-by-species quotas, one of the most fundamental changes the COT and FAO had recommended and one the SC repeatedly had called for as far back as the early 1950s, seemed no closer to being realized in 1970 than it had been in 1960. As before, the governments involved in Antarctic whaling were not prepared to accept a system they believed would present more problems for their industries, regardless of the obstacles a single BWU quota posed for the conservation of the remaining stocks.

5

SCIENTIFIC UNCERTAINTY AND THE EVOLUTION OF THE SUPERWHALE

The IWC's adoption of a moratorium on commercial whaling in 1982 was a major turning point in the commission's treatment of scientific uncertainty. By adopting the moratorium, the commission, for the first time in its history, interpreted and used scientific uncertainty as a basis for ceasing all commercial whaling. The IWC's adoption of the moratorium also effectively established the precautionary principle as its guide in policy making—a sea change in IWC direction that helped set the stage for a period when environmental interests increasingly would dominate commission policy.

As discussed in chapters 2, 3, and 4, the treatment of scientific advice and the uncertainties it entailed in the IWC from the 1950s through to the late 1970s was largely determined by how much the advice was seen to assist in the continued exploitation of whale stocks. This standard was initially defined by the need to continue hunting the maximum number of whales and then later, in the 1970s, by the need to limit hunting in order to prevent further depletion of stocks.

In this context, uncertainty about the need to catch fewer whales was clearly an important issue, since the costs of reducing catches would have been considerable for both the whaling industries and consumers. Scientific advice urging reductions in the Antarctic quota offered little in the way of utility to the five Antarctic whaling nations in terms of their criterion I priorities (i.e., recouping investment and remaining profitable), and also was in direct conflict with the internationally established criterion II need for more edible fats. As a result, recommendations for lower quotas were questioned, and ultimately rejected, on the grounds of uncertainty.

Therefore, it seems likely that any level of uncertainty concerning the

SC's advice, no matter how small, would have been invoked by the whaling countries and their industry representatives in order to avoid the financial repercussions of lower quotas. However, the industry's decline in the 1960s—caused by the shrinking of both the international demand for whale products and, most importantly, the number of whales—altered the priorities of many IWC members concerning criteria I and II. As a consequence, members also changed their perceptions of the uncertainty that had haunted those scientists calling for reduced catches and better management.[1]

The goal of keeping the whaling industry operational and able to provide whale products for consumers has continued to the present time. Indeed this still largely shapes the remaining whaling industries' perceptions and treatment of scientific uncertainty (albeit in an entirely different way). But demand for whale-based products has significantly declined and, as a result, profitable exploitation is no longer a commonly shared objective within the IWC and has not been for some time.

From the early 1960s onward, when the number of countries actively participating in commercial hunting began to fall, the economic and financial considerations that had driven IWC policies gradually became less relevant to an increasing number of the commission's members. This declining relevance led to a proportionately increasing level of commitment among the IWC membership to the views of the IWC scientists. As we have seen, many IWC scientists had been calling for reduced catches for some time, with the notable exceptions of scientists from Japan and Russia where whaling was still being actively pursued on a large scale in both the Antarctic and the North Pacific.

Once there were no longer any pressing domestic or economic reasons for governments to question the need for a more conservative approach to management (by raising problems related to uncertainty), it became easier for those members to alter their perception of uncertainty. Instead of using uncertainty to argue against reductions, governments began using it to argue in favor of smaller catches—particularly in light of the mounting indications that many stocks had been heavily overexploited. The rapidly declining catches of the late 1950s and early 1960s, in effect, had encouraged the belief that science aimed at conserving stocks was now very relevant if whales and whaling were to survive.

The decline in whaling's economic importance during the 1960s and in particular its limited cultural relevance among the IWC's predominantly Western membership essentially meant that research arguing in favor of smaller and better regulated hunts was no longer likely to conflict with any already established needs in most member states as described by criterion II (i.e., the importance of whaling for many consumers ended with whale oil's replacement by vegetable oils and the replacement of other products such as baleen and sperm oil with synthetic substitutes). This declining relevance of whaling set the stage for a short period in which scientific advice was privileged in the IWC. Significant progress in research and management methods was made as a result of fewer members invoking uncertainty in order to oppose scientific advice. Concurrently, the early 1970s saw an emerging awareness of environmental issues; the first calls for a moratorium on whaling came at the 1972 UN Conference of the Human Environment in Stockholm. These developments were redefining the criterion I and II priorities of a growing number of IWC members—a trend that would continue to alter the IWC's use and interpretation of scientific uncertainty for some time to come.

As awareness of the scope of environmental damage increased and the political value of supporting action to prevent further damage also increased in proportion to the growing number and influence of environmental nongovernmental organizations, an entirely new set of opportunities in relation to cetaceans began to emerge: in particular, the opportunity to exploit whales as an environmental symbol, thereby demonstrating one's commitment to protecting the environment.

In terms of criterion I, most Western societies in the post-Stockholm period turned to the goal (and utility) of protecting the environment (since whaling in the West had all but disappeared). As a result, scientific advice in the IWC began to be judged by the extent to which it achieved this objective. Criterion II barriers to this trend, in the form of consumer demand for whale products, were fast diminishing with the declining interest in whaling and were being replaced with increasing demand for more environmentally responsible policies on the part of governments and business. This sea change presaged a period that would not only generate increasing political support for the moratorium's adoption, but also would establish a major criterion II obstacle to the implementation of any scientific research sanctioning a return to commercial whaling: the superwhale.

SCIENTIFIC ADVICE IN THE 1970S: MORE EQUALS LESS

By the close of 1972, the IWC was operating in an entirely different environment—one that bore little resemblance to the earlier years of the commission's existence. Of the five Antarctic whaling nations that largely dictated IWC policy throughout the 1950s and much of the 1960s, only the Soviet Union and Japan continued Antarctic hunting into the 1970s. Pelagic whaling was no longer of any interest to the three former Antarctic nations—Norway, the United Kingdom, and the Netherlands—nor to anyone else in the commission bar the Japanese and Soviets. The Netherlands, following New Zealand's lead a year earlier, even went so far as to resign from the IWC.[2]

Also absent by this stage were the debates over whether quotas for the Antarctic stocks should be reduced—a reoccurring theme in the IWC that, more than any other, had defined the commission's management in earlier years. Quota debates by the close of the 1960s were instead characterized by disagreement over how much catches needed to be reduced, since the issue of whether reductions were needed had been resolved by both the collapse of Antarctic whaling and also by the findings of the Committee of Three and the Committee of Four. By far the biggest changes to date concerning the IWC, its management policies, and future direction, however, occurred in 1972. Of these, the single most important development was an unprecedented proposal from the U.S. delegation for a ten-year global moratorium on commercial whaling.[3]

The IWC's 1972 meeting in London followed the UN Conference on the Human Environment held in Stockholm earlier that year. The IWC meeting featured an address from Maurice Strong, the secretary general of the UN conference, in which he outlined the conference's resolutions on cetacean management. The Stockholm conference had recommended "the strengthening of the International Whaling Commission, increasing international research efforts, and ... an agreement involving all governments concerned in a ten-year moratorium on commercial whaling."[4] The U.S. delegation responded by proposing in the Technical Committee, with support from the United Kingdom, "a motion that the schedule for 1973 be amended in every case where a numerical quota appears to substitute the numeral 'o' for all such numerical quotas."[5]

The U.S. government's push for a temporary end to all commercial whaling was among several steps it only recently had taken to provide greater protection for the great whales, including several unilaterally imposed regulations passed by Congress.[6] The U.S. proposal was based on the assertion that the levels of uncertainty and knowledge gaps in the IWC's existing research and regulations were too significant to justify any further commercial hunting. The U.S. delegation argued that "the state of knowledge of the whale stocks was so inadequate that it was only common prudence to suspend whaling; this was necessary so that scientific efforts could be redoubled and new research techniques developed."[7]

The U.S. position was a clear reinterpretation of what scientific uncertainty should mean in terms of wildlife resource management, one that took the COT's earlier recommendations for conservation to their management extreme. The moratorium proposal was defeated in both the TC and later in the plenary (four delegations voted in favor, six against, and four abstained). But its introduction and minority support[8] was enough to bring on a debate that would dramatically alter the IWC's management priorities and also fundamentally question the SC's ability to deal with the uncertainty issues facing the commission.

Divisions within the SC over the status of the Antarctic stocks, in particular the fin whale, were no closer to resolution at the 1972 meeting; the Japanese and Soviet scientists continued to support catches that were larger than what others in the SC were prepared to accept.[9] Sufficiently reliable data for more accurate and convincing estimates, however, remained out of the SC's reach. In addition to the ongoing disputes over the status of the Antarctic stocks, fears were increasing that fin, sei, and more recently sperm whale stocks in the North Pacific and elsewhere were coming under threat, adding further weight to a growing body of opinion outside the IWC that U.S. concerns for the future of the great whales were probably justified.

Indeed, the UN-sponsored demand for a temporary end to commercial whaling clearly had demonstrated the extent to which the IWC's policies were now attracting international attention. Thus, in lieu of the IWC implementing a management program broadly accepted as reliable and safe, it appears to have been inevitable that pressure for the moratorium's adoption—both within and outside the commission—would continue to grow.

It is, however, ironic that calls for a moratorium on commercial whaling began at almost precisely the same time that the SC finally succeeded in convincing the IWC to abandon the BWU and adopt species-by-species quotas. At the 1972 meeting, provision for the BWU quota was deleted from the schedule and all quotas were subsequently set on a species-by-species basis for Antarctic baleen whale stocks, a move that allowed maximum sustainable yield to become the commission's guiding management principle in both theory and practice. This landmark decision by the IWC followed its adoption of North Pacific quotas for baleen whales (fin, sei, and Bryde's whales) in 1970 and also for sperm whales in 1971, marking the first time that quotas for stocks outside the Antarctic were incorporated into the schedule.

The other milestone achieved by the IWC at its 1972 meeting was the long-awaited implementation in all regions of the International Observer Scheme. The IOS had first been proposed by Norway in 1955, but actual agreement on the scheme by the then five Antarctic nations was not reached until 1963. Even then its implementation was stalled by the Soviets, who linked cooperation on the IOS to a satisfactory resolution of the deadlocks concerning national quotas.[10] Agreement on the IOS finally was reached at the IWC's 1971 meeting in Washington; the scheme would be implemented in the North Pacific, the Antarctic, the North Atlantic, and at all Southern Hemisphere land stations for the 1971–72 season. But implementation in the Antarctic again was postponed—this time only until the following season—due to "practical difficulties."[11] In its final resolution of the IOS, the commission addressed the costs involved and the selection and placement of observers, issues that had caused considerable difficulties and delays: "There shall be received such observers as the member countries may arrange to place on factory ships and land stations or groups of land stations of other member countries. The observers shall be appointed by the Commission and paid by the government nominating them.[12]

Thus from 1972, the IOS was in force in all its designated regions—a move that had long been needed in order to allow some verification of the actual catches made. But the full importance of the IOS scheme would not be realized until more than two decades later, when detailed evidence surfaced showing that Soviet fleets had falsified catch reports throughout the 1950s and 1960s. Some in the commission had suspected this but could not substantiate their concerns. Needless to say, the admission by Russian scientists forced the SC to revise its catch history data for most species.[13]

Given the substantial improvements in the IWC's approach to management that had occurred by 1972, in particular the abandonment of the BWU system, the joint U.S./UK push for a blanket moratorium covering all stocks in all regions was not well-received by the SC, which had long been making its own push for species-by-species management. In response, the SC rejected the moratorium and called instead for a program of more intensive research that would later become known as the International Decade of Cetacean Research:

> The Committee agreed that a blanket moratorium cannot be justified scientifically. A blanket moratorium is in the same category as a blue whale unit quota, in that they are both attempts to regulate several stocks as one group whereas prudent management requires regulation of the stocks indi-

vidually. The committee noted that the absence of commercial catching operations would make it impossible to obtain certain kinds of information which are essential for continuing assessment of whale stocks. There is in fact a need for a substantial increase in all kinds of research activity related to whales. Therefore the Committee recommends that, instead of a moratorium, support should be sought for a decade of intensified research on cetaceans, particularly as regards problems relevant to their conservation. Such a program should proceed in parallel with further development by the Commission of the policy of bringing catch restrictions into line with the best available knowledge of the state of stocks.[14]

So while the moratorium proposal was rejected by the plenary on the advice of the SC, the committee's counterproposal for a decade of "intensified research on cetaceans" was supported by the commission. General support also was given to the measures put forward by the United States to address the UN conference's resolution on the need for "strengthening" of the IWC. This included plans for expanding the commission's staff and also its membership, especially among commercial whaling nations operating outside the convention.[15]

The United States repeated its call for a ten-year moratorium on commercial whaling at the 1973 and 1974 IWC meetings, but on both occasions the SC maintained its opposition while again stressing the need for the commission to continue managing stocks on an individual basis. But with the moratorium's proponents increasingly emphasizing the problems posed by the many uncertainties that characterized the SC's management advice, pressure for more convincing and unequivocal estimates and quotas began to mount.

One problem facing the IWC scientists in their bid to create a safer and more reliable management system was the even larger amount of data and analysis required to set MSY-based quotas for individual stocks, particularly since quotas now needed to be set for a larger number of species in both the Southern Hemisphere and the North Pacific. In addition to providing catch estimates for the fin and sei stocks in the Antarctic, which had occupied most of the SC's time during the late 1960s, and monitoring the condition of the protected stocks, scientists now also needed to calculate quotas for minke, Bryde's, and sperm whale stocks, since catches of these species had increased significantly with the ongoing decline in fin whale catches.

The SC's efforts were further complicated by the disagreements over stock estimates that continued to divide its members. The SC used increasingly sophisticated MSY and stock-by-stock management methods (including plans for Antarctic baleen species quotas to be subdivided into stocks reflecting "the natural subdivisions in the distribution and genetic composition of the stocks"[16]). But the more certainty scientists tried to provide, the more uncertainty they encountered due to a lack of basic data that, as a consequence, then required further assumptions concerning fundamentally important biological parameters.

Thus, with the exception of the numbers given for sei and Bryde's whales (both species were grouped under one quota), much of the SC's analyses and estimates were in dispute, stemming from the disparate views of the Japanese and Soviet scientists on the one hand and the remainder of the committee's scientists on the other who tended to take a more conservative view. Prior to this point, when even suggestion of a global moratorium had been all but impossible, failure to provide unanimous advice on stock estimates and quotas invariably had resulted in the Antarctic nations using uncertainty to argue for higher quotas. However, with support for zero catches now gaining significant support, any disagreement within the SC enabled uncertainty to be used by the pro-moratorium nations to support their calls for an end to commercial whaling.

The interpretation and treatment of scientific uncertainty in both the plenary and the SC was indeed at something of a crossroads during the early 1970s. From 1972 onward, uncertainty issues were simultaneously understood to mean different things by different people due to the division of interests that now existed within the commission. As a consequence, the IWC was moving toward a situation where more and more policy debates in the plenary gradually focused on a choice between two likely management outcomes: setting quotas at or near a level that would satisfy the Japanese and Soviet industries or setting quotas at zero.

But by as early as 1973, the direction the IWC finally would take in this regard was becoming clear. The policy pendulum had already swung away from an extreme that had exclusively favored industry interests to pause briefly at a point where the needs of both the industry and the resource were being considered. As the political momentum behind the pendulum's initial shift continued to gather force, however, it soon was on its way toward the alternate extreme: a global moratorium on hunting.

The extent and nature of the split between the IWC's shrinking fellowship of active whaling nations (including the only two surviving pelagic whaling nations, Japan and the Soviet Union) and its slowly expanding clique of nonwhaling nations (many of whom, most notably the United States and the United Kingdom, either had given up what remained of their former whaling interests only recently or were in the process of doing so) were well illustrated at the IWC's 1973 meeting, again held in London. The U.S. delegation picked up where it had left off the previous year by once more proposing the adoption of a ten-year moratorium on commercial whaling, after having gained simple majority support for its moratorium resolution in the TC. Again, the U.S. proposal failed to gain the necessary three-quarters majority support, with the SC reiterating its objections to the proposal from the previous year. What was remarkable, however, was that for the first time in IWC history a simple majority of members—eight in favor, five against, with one abstention—had voted in the plenary in support of a global ban on all commercial whaling.[17]

The management priorities of many in the IWC clearly were undergoing dramatic change and already had altered markedly among several members in the short space of only a year. Furthermore, not only was the U.S. government reinforcing its opposition to the commercial interests of the relatively few IWC members still committed to whaling, pelagic or otherwise, it now was attempting to do so by directly questioning the SC's ability to advise on management proposals.[18] After again basing the need for a moratorium on the belief that "knowledge was inadequate to protect the species"—a point supported in part by four U.S. scientists now advocating protection of Antarctic fin whales on similar grounds[19]—the U.S. commissioner went on to state that his delegation also "had reservations in respect of the recommendations of the Scientific Committee."[20] The short period of *relative* calm and general willingness to adhere to the SC's advice that had existed in the IWC since the late 1960s was, it seemed, at an end.

In addition to the prevailing dispute over the Antarctic fin whale stocks, questions concerning the MSY of Antarctic minke whales contributed to the political pressure on the SC. The repeated calls for a moratorium, coupled with the disputes over Antarctic quotas and area subdivisions, already had resulted in Japanese and Soviet threats to discontinue the IOS in the Antarctic.[21] At the 1973 meeting both governments upped the ante further by returning to their past practice of using objections to veto changes to the schedule if those changes disadvantaged their industries. On this occasion, the measures objected to by the Japanese and Soviets involved plans to phase out Antarctic fin whale catches and also a decision against Japan's proposal to increase the existing Antarctic minke whale quota from 5,000 to 8,000.

In response to the steadily falling catch of Antarctic fin whales,[22] the commission voted to reduce fin whale catches in the Antarctic for the 1973–74 season and also to cease all catching of the stock by the 1976–77 season. The Japanese and Soviet commissioners accepted the quota reduction but opposed the phaseout of catches (the Soviets, however, initially expressed support for the phaseout in the plenary), while also objecting to the commission's rejection of Japan's demand for a take of 8,000 minke whales.

> At its Twenty-Fifth Meeting, the Commission amended the Schedule to provide for catch limits for baleen whales in the Antarctic for the 1973/74 season. The limits were set by species as follows: 1,450 fin whales, 4,500 sei and Bryde's whales combined and 5,000 minke whales.
>
> It was decided that the taking of fin whales in the Antarctic should cease not later than June 30, 1976. This was opposed by the Japanese and Soviet delegations. The Japanese Government objected under Article V, of the Convention to the Amendment to the Schedule to implement this decision and it is not therefore binding on that government. The Japanese Government and the Government of the USSR also objected to the amendment setting the catch limit of 5,000 minke whales in 1973/74 on the grounds that it did not reflect the actual state of the stock and the amendment to the Schedule is not binding on these countries. The Commission was informed that the two countries had each agreed to limit its [sic] catches to 4000 minke whales.[23]

The dispute within the SC over the Antarctic minke whale stocks illustrates the difficulties its members faced in the absence of sufficient data, and also the extent to which efforts aimed at reaching agreement were handicapped by growing discord within the committee over how uncertainty issues should be handled. Chapman, the committee's chairman and former member of the COT, rejected a MSY estimate for minke whales presented by Japanese scientist Seiji Ohsumi, saying that Ohsumi's estimate of 12,230 animals was "much too large and that a more reasonable estimate on the basis of analogy with other species is about 5,000."[24]

Ohsumi countered by insisting "that the population size was reasonably revised on the basis of improved sighting coverage in 1972/73 and therefore Chapman's estimate is unreasonably too low even in the light of the last report."²⁵ Other scientists in the committee, in particular South African scientist Peter Best, urged caution, noting both the provisional nature of the new and "radically revised" Japanese estimate of the initial population (299,999 versus 150,000 in 1972) and that "both estimates of MSY are based on assumptions regarding the values of critical parameters and that there is as yet no sound basis for determining which is the more accurate."²⁶ In its advice to the commission, the SC reflected Best's assessment by admitting that "it had no sound basis for determining which of the estimates [i.e., 5,000 or 12,230] was the more accurate" and the SC "emphasized the importance of a conservative approach."²⁷

For the Japanese and Soviet delegations, however, a "conservative approach" meant a catch of 8,000 minke whales (Japan had previously pressed for 8,500 in the TC), with the Japanese arguing that the SC agreed there was a "large surplus" above the MSY level.²⁸ But the rest of the commission remained unconvinced and opted for the lower estimate provided by Chapman. Antarctic whaling interests and profitability clearly were no longer able to dominate IWC policy as they had prior to and during the Antarctic collapse.

As a result, efforts by Japan's scientists to continue arguing uncertainty issues in the industry's favor were more broadly resisted in both the SC and the plenary.²⁹ All but two of the commission's member governments—in spite of the continuing threat of objections—appear to have fundamentally changed their interpretation of uncertainty issues, as they now were prepared to give the benefit of the doubt to the resource. This represented a clear break from the IWC's past policy decisions that gave priority only to the immediate economic needs of the industry.

The Japanese and Soviet decision to return to the tactic of objecting to unfavorable changes to the schedule resulted in a catch of 7,713 minke whales during the 1973–74 Antarctic season (Japan 3,713, Soviet Union 4,000) as opposed to the 5,745 animals taken in the previous season. Their insistence on a higher catch also provoked an unprecedented display of international opposition—adding a new dimension to the already dramatically altered political environment confronting the two remaining Antarctic nations. In addition to the increasing media attention to the IWC meetings, a growing number of environmental and animal welfare groups also were attending the annual meetings, generating further pressure to adopt a more precautionary approach (in favor of the stocks).

The first such NGO to be listed as an observer in the commission's annual reports was the International Society for the Protection of Animals, which sent a representative to the IWC's 1963 London meeting. By 1965, the number had grown to three with the addition of the World Wildlife Fund and the Fauna Preservation Society, but showed no further significant increase until 1972 when the number of groups doubled to six. Then, by the 1974 meeting, following the Japanese and Soviet objections, the number of environmental/animal welfare NGOs stood at eight, after which attendance among NGOs of various persuasions increased rapidly.

At the 1982 meeting in Brighton—the year the moratorium was adopted into the schedule—more than fifty NGOs were in attendance, with the overwhelming majority expressing strong support for an end to commercial whaling and many also supporting a ban on aboriginal whaling.[30] Importantly, the Soviet and Japanese refusals to accept stronger conservation measures at the 1973 meeting, and the reactions that followed, had created something of a turning point in the IWC, with repercussions that would be felt for some time to come. According to James E. Scarff,

> That action [i.e., the Japanese and Soviet objections] triggered a wave of international protest and caused several American conservation groups to organise a boycott of all Japanese and Soviet imports. The boycott was to last until those countries had publicly committed themselves to international cooperation and whale conservation. By the spring of 1974, 21 American conservation, humane, and environmental groups, with more than five million members, had pledged to support the boycott. Although it is difficult to evaluate the effect of the boycott, at least one observer at the 1974 meeting felt that its impact had been profound.[31]

By 1974, the political environment within the IWC had become more complicated than ever due to both the greatly increased level of public attention and the growing concerns over the many uncertainties about the status of the exploited stocks. Environmental problems and threats were becoming issues of broad public concern in a number of the IWC member countries, in particular the United States. Governments and companies thus risked significant political and economic backlash if they ignored the groundswell of public opinion being generated by an increasingly influential environmental NGO lobby.

With political pressure for greater stock protection mounting as criticism of the uncertainties in the SC's advice increased,[32] the IWC looked as if it might return to its earlier periods of crisis and threats of resignation. A compromise came in the form of a new and more cohesive management strategy that became known as the New Management Procedure.³³ For a short time this initiative provided some hope that the IWC could build upon its recent achievements by maintaining enough of a cooperative atmosphere among its members to realize further progress. But as William Aron has observed, the commission's expectations for the NMP as a cure for its political divisions and management problems soon would prove ill founded:

> By the mid-1970s the pressures from the environmental community caused real changes in the IWC. The elimination of the Blue Whale Unit, forcing the management of whales by species and stocks and the implementation of the New Management Procedures, provided a strong and conservative underpinning for whale management. An era of successful resource management was anticipated—a prediction that was foolishly optimistic.³⁴

At the IWC's 1974 meeting, Australia, seconded by Denmark, proposed that the commission adopt the NMP in response to the mounting tension between the commission's whaling and nonwhaling members over the moratorium issue. In making its proposal, the Australian delegation referred to "the need to preserve and enhance whale stocks as a resource for future use and taking into consideration the interests of consumers of whale products and the whaling industry as required by the International Convention on Whaling."³⁵ This new management approach, the NMP, classified all stocks into one of three categories on the advice of the SC. The Chairman's Report described the categories and their initial definitions:

> (i) *Initial Management Stocks* which may be reduced in a controlled manner to achieve MSY levels or optimum levels as these are determined.
>
> (ii) *Sustained Management Stocks* which should be maintained at or near MSY levels and then at optimum levels as these are determined.
>
> (iii) *Protection Stocks* which are below the level of Sustained Management Stocks and should be fully protected.
>
> The Committee would define stocks for this purpose as the units which can be most effectively managed individually.³⁶

The Australian proposal for the NMP—primarily the conception of Australian scientist and former COT member Kay Radway Allen—was a milestone in the IWC's history of attempting to manage whale stocks. In addition to offering a possible solution to the escalating conflict over stock management by attempting to take at least some of the uncertainty into account, it also significantly increased the influence of the SC in the setting of quotas. The NMP proposal and its implementation deadline for the 1975–76 season received broad support in the plenary and was adopted by a majority vote. Even the U.S. government—the most outspoken critic of the current situation and the uncertainties it involved—seemed satisfied enough to endorse the resolution, thereby postponing, for the moment at least, its insistence on a global moratorium[37]. The U.S. commissioner explained the U.S. position, saying that while his government "still supported a ten-year moratorium [the U.S.] had voted for the resolution [i.e., on the NMP] because it felt it represented a significant step forward in the management of the world's whales."[38]

The SC held a special meeting later that year (1974) at La Jolla to determine the definitions and rules needed for the NMP's implementation. The committee presented the following designation of stocks under the NMP at the 1975 IWC meeting, as summarized by the IWC's scientific editor, Greg Donovan:

(1) Initial Management Stocks: those over 72 percent of the original stock size; catch limits are always set below MSY to ensure that the stock is not reduced below 60 percent, the MSY level.

(2) Sustained Management Stocks: those between 54–72 percent of the original stock size; catch limits are again set below MSY, the degree below depending on how far below the MSY level the stock is.

(3) Protection Stocks: those below 54 percent of the original stock size; no catches are allowed on such stocks.[39]

The NMP's adoption at the 1975 meeting and implementation in the following season resulted in quota reductions for baleen catches in both the Southern Hemisphere and North Pacific, particularly for the remaining unprotected fin whale stocks. Quotas for fin and minke stocks in the North Atlantic also were set for the first time.[40] The SC, could now set quotas for most of the world's exploited stocks with minimal opposition to its recommendations from the commissioners, since the NMP required that catches be set according to the biological limits of the stocks rather than the economic criteria of the industries.

In an attempt to deal with the uncertainties and knowledge gaps in the data used by the scientists, the NMP was equipped with several somewhat simple safeguards designed to allow for a greater margin of error in estimates for the benefit of the stocks. These safeguards mostly were based on what became known as "the 10 percent rule": an arbitrarily determined allowance for error that provided a safety margin of 10 percent in the calculation of both MSY and the level below MSY at which a stock should be designated as a Protection Stock. In the case of Sustained Management Stocks and Initial Management Stocks, for example, no more than 90 percent of the estimated MSY could be taken. In the case of determining protection status, stocks believed to be at 10 percent or more below their assumed MSY level (60 percent for baleen whales) were classified as Protection Stocks.[41]

At the 1976 meeting, the IWC took further steps toward conservation of stocks with the protection of fin whales in the Southern Hemisphere and North Pacific, in addition to the establishment of quotas for minke whales in the North Pacific and for North Atlantic stocks of sei and sperm whales. As of 1976, not only were all stocks being managed under quotas for the first time, they also were being managed in accordance with the "conservation first" approach that underpinned the principles of the NMP.[42] Furthermore, in accordance with the recommendations of the UN's 1972 Stockholm recommendations, the IWC administration was strengthened just prior to the 1976 meeting by the appointment of a scientist, Dr. Ray Gambell, as full-time secretary to the commission. In the same year, the commission also accepted from the SC a program of research to begin the earlier called for International Decade of Cetacean Research.[43] These important changes, however, would soon be overshadowed by the emergence of serious problems with the NMP and the data it needed, a development that revived the moratorium debate.

The NMP's main shortcoming—a requirement for precise biological data and population estimates that were neither available nor considered achievable[44]—created further divisions within the SC and the plenary. Reliability of the NMP increasingly came under question, particularly from the United States and a growing number of other governments that were, with the support of anti-whaling NGOs, once again calling for the adoption of a ten-year global moratorium. The biggest flaw in the NMP was that it had not gone far enough in trying to account for the burgeoning array of unknowns that faced the SC in its task of managing dozens of stocks all over the world. As a consequence, it became relatively easy for those supporting the moratorium to highlight the many uncertainties involved in the NMP's application. Critics then used this uncertainty in support of the growing calls for a moratorium, which had reemerged by the late 1970s. According to Aron,

> In trying to implement the terms of the NMP, it became obvious that the procedure was flawed, particularly because of the paucity of critical data and serious uncertainties about the classification scheme. While it provided a scientific rationale that allowed selective moratoriums to be implemented, it did not, in the view of some, provide assurance that the harvested stocks were being taken at catch levels that would ensure their long-term sustainability.
>
> The scientific uncertainties in implementing the NMP coincided with a period of rapid growth in the membership of the commission, almost exclusively by nations committed to a ban on commercial whaling. The original fourteen-member commission swelled to thirty-three members by 1981 (thirty-nine by 1996). The domination of the commission by members opposed to commercial whaling allowed the uncertainties of the Scientific Committee to be translated into a ban on commercial whaling.[45]

Thus, the 1970s, a period of major improvements in both the management methods used by the SC and the functioning of the committee itself, saw the influence of scientific advice in the IWC grow but then quickly decline. SC members were confronted with the limits that scientific uncertainty imposed upon them and their attempts to better understand the responses of cetacean populations to exploitation. The important issue was not the mere existence of uncertainty and knowledge gaps—these problems were all too familiar to many in the SC—but rather the question of what the uncertainties should be taken to mean in relation to management policy. The uncertainty issues in the late 1970s were largely the same as those that had divided and weakened the commission two decades earlier. The difference now was the political environment in which these uncertainties were being received and interpreted.

THE FALL OF THE NMP AND THE MORATORIUM'S ADOPTION

The central issue in the moratorium debate was whether sufficient information on the status of whale stocks existed to allow continuing exploitation without reducing stocks to levels approaching extinction. As the extensive information on whale populations required by the NMP was unavailable, and the procedure itself insufficiently accounted for errors in stock size and catch quota estimates, many in the commission's increasingly nonwhaling membership worried about a repeat of the overhunting that had characterized the IWC's performance in the 1950s.

The SC, however, had expressed serious reservations over the a blanket moratorium in response to the Stockholm conference on the grounds that since not all species were endangered, there was no reason to place all stocks under protection.[46] The SC had dedicated a great deal of effort in the 1960s and early 1970s to establish, on the basis of the work done by the COF, the principle that whale species should be managed individually rather than collectively. And for this reason, the moratorium was without justification in the eyes of many IWC scientists, as all endangered stocks were already under IWC protection.

But although many members of the SC continued to oppose a blanket moratorium, by the late 1970s the committee's strong opposition to the moratorium was being weakened by an emerging minority view; some scientists within the committee, in particular Holt and later Chapman, had begun advocating an even more precautionary approach, ostensibly because of the NMP's shortcomings and the uncertainties involved in the stock assessments it used.

By 1979, divisions within the SC over the "adequacy" of the NMP had become manifest. The committee's 1979 report outlined several areas where uncertainty issues were of particular concern and mentioned the disagreement between "some" and "other" members over both the risks involved with the NMP's use and also the likelihood of research being hindered rather than aided by the imposition of a moratorium.[47] The report explained,

> In its last report . . . the Committee expressed the opinion that "the present management procedure does safeguard whale species and identified major stocks for future generations at a very small risk." Some members of the Committee still believed this to be a true statement. Other members expressed some doubts in the light of difficulties met in assessing the state of, for example, stocks of sperm whales and Bryde's whales, and of the fact that whaling on a considerable number of stocks was now, or was proposed to be regulated under interim measures. It was recognised that the Commission was trying to improve the management procedures. Some members thought, however, that such improvement would be slow and that

continued whaling, at least on some stocks and species, at current levels could indeed involve considerable risk if not of extinction then surely of depletion to levels at which they would need protection even under the present rules. Others stressed that as recognised in the Committee's last report, although the critical minimum population sizes have not yet been identified, a stock is very unlikely to be threatened with extinction before the need to reclassify it becomes evident.[48]

Holt, an FAO observer at IWC meetings and a former member of the COT (and COF), became one of the most outspoken critics of the SC from the mid-1970s onward, particularly in relation to the SC's handling of the uncertainty issues that he believed made the NMP unreliable:

> In the 1970s, the regulation of whaling in the Antarctic, or anywhere in the world, had not really been resolved. And the commission had to respond to the United Nations call for a moratorium on commercial whaling and did so by producing . . . what they call the New Management Procedure. This is based on scientific modelling. But by the end of the 1970s it was pretty clear that the science was not up to properly applying that procedure—it had broken down for lack of data, lack of evidence, and uncertainty in the whole process. That led to further pressure for a moratorium.[49]

Indeed, Holt made his concerns clear in numerous reports, papers, and minority statements that he submitted to the committee during this period.[50] A number of his contemporaries saw him as an important figure in the use of uncertainty arguments concerning the NMP's shortcomings as a basis for the moratorium's adoption. According to Gambell,

> The chief proponents of uncertainty to devalue the work of the Scientific Committee in the 1970s were Sidney Holt, John Beddington, Bill de la Mare and Justin Cooke.
> Holt was there in a number of guises—representing the Seychelles (together with Lyall Watson), or an NGO organisation. John Beddington is a British mathematician who quickly realised the damage his position was causing to his scientific credibility, and so pulled out at an early stage. This group was subsequently joined by a Dutch scientist, Kees Lankaster. Bill de la Mare worked originally for Greenpeace, and then went to the Australian government service, and the fourth member, Justin Cooke (another Brit), represented an NGO and then IUCN [the World Conser-

vation Union]. This group was probably motivated by the anti-whaling philosophy of the time, but had the scientific authority to undermine the best efforts of the rest of the SC, which of course was seized on by the anti-whaling Commission members.

Doug Chapman from the USA was a member of the Committee of Three (as was Holt), and he was held in high regard as a mathematician. So it came as something of a surprise to all of us when he and Mike Tillman, who had both been chairmen of the SC, came out on behalf of the US government against accepting the work of the SC as a basis for . . . commercial whaling.[51]

By the end of 1979, the divisions within the SC over the uncertainties and risks involved with the NMP brought about further restrictions on commercial whaling and also renewed calls for the moratorium's adoption. At the IWC's 1979 meeting in London, the commission voted with the required three-quarters majority to accept the Seychelles' proposal for the establishment of an Indian Ocean Sanctuary and ban all pelagic catches excluding minke whales.[52] Moratoria on all but a very limited amount of land-based and pelagic whaling had been adopted, and it seemed likely that the increasing dissatisfaction with the NMP would soon push the commission toward adoption of a global moratorium. The United States—with strong support from like-minded governments, now also including Australia, which had proposed the NMP only four years earlier[53]—again was leading the charge to end all commercial whaling. In his opening statement at the IWC's 1979 meeting, U.S. commissioner Richard Frank told the commission,

> This 31st annual meeting of the Commission is one of the most critical in our history. Will we come to grips in a forceful manner with the inadequacies of the current system for regulating commercial whaling and truly endeavor to fulfill our responsibilities to conserve and manage the great whales? Or, will we continue on a path which subjects whale stocks to unacceptable risks?
>
> The fundamental issues before us are the various proposals for a moratorium on commercial whaling. We attach the highest importance to these proposals. We are distributing today a letter from the President of the United States urging the Commission to adopt an indefinite moratorium, thereby at last taking effective action to ensure the survival of the great whales.
>
> The United States has long supported the concept of a moratorium on

commercial whaling. Since 1975, when the New Management Procedure was adopted with the support of the United States, we have not pressed the moratorium issue in this forum. However, it has become increasingly clear that our management system is not functioning as adequately as we believe necessary. The time has thus come for a change.[54]

The IWC's few remaining whaling countries, with the earlier support of Canada, argued strenuously against the moratorium, citing its lack of scientific support and the FAO statement made to that effect in 1982.[55] But, due to the large increase of nonwhaling members in the commission during the late 1970s and growing doubts within the SC over the reliability of the data required by the NMP, pro-whaling governments were unable to prevent the moratorium from gaining the required three-quarters majority for its adoption at the 1982 meeting.[56] The whaling governments were not the only ones to oppose the moratorium on the grounds it was based more on politics than on compelling scientific arguments. Switzerland also questioned the moratorium's scientific basis and abstained from voting.[57]

At the 1982 meeting, the Saint Lucia commissioner, responding to an FAO statement that also questioned the moratorium's scientific basis, asked whether "satisfactory scientific evidence" existed to support the view that catch limits were "within the productive capacity of the stock and are indefinitely sustainable." Two years later the SC responded that while "some members believed . . . there is little or no satisfactory evidence" that stocks could indefinitely support recent catch limits (due to the lack of information on population sizes and dynamics), "many members believed that catch levels (not necessarily at a constant rate) could be kept within the productive limits of a species, given an understanding of population trends and if proper procedures were in place for changing the catch levels as needed."[58]

Significantly, Holt and Chapman, who expressed some support for a "negotiated interim cessation of commercial whaling . . . to ensure the future productivity of whale resources" also had been members of the COT.[59] Their views on the moratorium, however, were not shared by Australian scientist Allen,[60] also a former COT member, or by John Gulland, the fourth scientist on what became the COF. Gulland described the moratorium as "hardly an important victory for conservation, even in the context of whaling. In a wider context it was closer to a setback."[61]

The governments of Japan, Norway, Peru, and the Soviet Union each lodged objections to the moratorium (Iceland and the Republic of Korea also opposed the moratorium but neither lodged an objection) within the required ninety-day period in accordance with the International Convention for the Regulation of Whaling. So they were not obliged to adhere to the imposition of zero catch limits.[62] Japan, however, later withdrew its objection and ceased commercial whaling in 1988 under threat of U.S. sanctions on Japanese fishing in U.S. waters. Norway and the Soviet Union also halted their commercial whaling operations in 1988 but did not withdraw their objections to the moratorium.[63] Thus, the complete pause in commercial whaling that occurred from 1988 on resulted more from U.S. political pressure than from scientific advice, especially given the controversy surrounding the moratorium's scientific basis and the division within the SC. As Andresen has noted, "the USA managed what the IWC had no means to accomplish; halt commercial whaling. Traditional bilateral diplomacy and power politics was the name of the game; not new scientific evidence."[64]

By separating politics and science in a way that suggests one is independent of the other, however, Andresen's observation fails to take into account the extent to which the U.S. government had been employing uncertainty to reinforce and justify its use of "traditional bilateral diplomacy and power politics." In fact, it seems clear that, through invoking scientific uncertainty in order to dismiss the SC's general (but not unanimous) opposition to the moratorium, the United States was demonstrating the extent to which uncertainty could be used to politicize scientific advice by treating it as either an ally or a foe, depending on the policy objectives at the time. Indeed, the issue of managing and interpreting scientific uncertainty had gradually grown in importance since the 1960s. By the late 1970s, scientific uncertainty again had become a powerful weapon, except this time it was a weapon that, unlike in the 1950s, was wielded in support of a proposal to bring whaling to a temporary end.

The commission's past use of scientific uncertainty to reject SC advice to lower catch quotas had successfully delayed the imposition of smaller catches but had also resulted in the overhunting of many of the great whale stocks, in particular the blue whale—a situation that eventually endangered both the larger whale species and pelagic whaling itself. Thus, a moratorium could now be argued for on the basis of a lower threshold for acceptable risk, which essentially translated into the view that scientific uncertainty should signal caution rather than business as usual. And so from 1982 onward, the precautionary principle effectively became the

IWC's guiding tenet in managing and interpreting scientific uncertainty. This watershed in commission policy provided the impetus for a comprehensive assessment of whale stocks, which in turn led to the creation of the NMP's replacement: the Revised Management Procedure (RMP).

THE PRECAUTIONARY PRINCIPLE AND THE DEVELOPMENT OF THE RMP

The IWC's creation of the Indian Ocean Sanctuary and banning of pelagic whaling of all species except the minke at the commission's 1979 meeting was a significant turning point for the commission. It reflected the extent to which anti-whaling sentiment already had grown in the IWC—in tandem with the IWC's steadily increasing nonwhaling membership—and also the dissatisfaction of many commissioners and some scientists with the NMP. By the early 1980s, anti-whaling governments and NGOs had made scientific uncertainty the preeminent issue in the commission and had all but formally established the precautionary principle as the basis of IWC policy. Unlike in the past, the majority of commissioners now demanded positive evidence that stocks were not endangered before catch limits could be set, pointing to the problems with the NMP and the uncertainty surrounding whale stock estimates as justification for this policy.

The increasing concern within the IWC over the level of uncertainty connected with stock estimates and the NMP's inability to manage uncertainty was underpinned by a growing international mood against whaling—propagated by various environmental NGOs throughout the 1970s—that had even begun to cast doubts over the commission's long-held assumption that sustainable whaling was a desirable activity. This signaled a radical change in the attitudes of many former whaling members towards whaling. On one level, the increasing acceptance of the precautionary principle within the IWC and the trend toward protectionist policies can be attributed to the shortcomings of the existing research and the resulting divisions in the SC. But on another level, it is also necessary to consider the changing political environment that led to the commission's more conservative approach to managing risk and to embracing the precautionary principle.

The moratorium and development of the RMP became necessary, in the eyes of many IWC members, precisely because of the way in which the IWC's altered political priorities had redefined the import and implications of scientific uncertainty as per the precautionary principle. In other words, by the late 1970s, the economic needs of the remaining whaling countries in the IWC had been replaced by a more broadly perceived need within the commission for greater protection against the overexploitation of whale stocks. Thus, for many IWC members, the previously recognized utility (criterion I) in hunting whales for commercial gain had been supplanted by a newly recognized utility in protecting them in order to appear more environmentally responsible. By protecting whales, governments enjoyed increased political support. Since whaling no longer offered any significant economic or political advantages in most Western societies, there was little opportunity cost involved in opposing it.

There were, however, considerable domestic political costs involved in supporting whaling for any government whose constituency (whether national or local) had adopted the protection of whales as a legitimate criterion II need (i.e., had decided that the disadvantages of whaling outweighed its benefits). By the end of the 1970s, many Western governments had realized that any further support of whaling would put votes in jeopardy but provide no tangible domestic benefits. And since opposing whaling would cost nothing and was only likely to enhance reelection prospects (i.e., political leaders and parties would appear committed to environmental protection), the choice on which position to adopt was probably a simple one for many IWC members.

This change in values and priorities was also changing the traditional interpretations among many IWC members of what "overexploitation" and "sustainable utilization" represented in terms of management issues and how they should be dealt with. The RMP, therefore, was also the product of a political response to the growing calls for greater environmental sensitivity rather than simply a "scientific" solution to the IWC's relatively new preoccupation with managing scientific uncertainty. The largely political questions of "how much risk is acceptable?" and "what should sustainable utilization mean?" have underpinned the IWC debate over managing scientific uncertainty. They are essential for understanding the RMP's development and also why it is yet to be implemented.

The precautionary principle played a crucial role in the 1982 moratorium. In the face of claims that a complete pause on all commercial whaling was not based on scientific evidence, the precautionary principle became the framework for the U.S.-led arguments against whaling. Its successful invocation by the IWC's majority of anti-whaling members effectively lowered the commission's tolerance of scientific uncertainty in whale management. Later it would provide the catalyst for the RMP as part of the overall comprehensive assessment of whale stocks, which the IWC called for in conjunction with the moratorium's adoption.

But, despite the growing anti-whaling mood within the commission, even the 1982 moratorium only represented a temporary pause until the concerns over scientific uncertainty were resolved. The issue of permanently banning commercial whaling had not yet formally been raised. This implied, at this stage at least, that the commission believed it possible to reduce scientific uncertainty to acceptable levels. Accompanying the moratorium's entry to the ICRW schedule was the following clause:

> This provision will be kept under review, based upon the best scientific advice, and by 1990 at the latest the Commission will undertake a comprehensive assessment of the effects of this decision [i.e., the imposition of zero catch limits] on whale stocks and consider modification of this provision and the establishment of other catch limits.[65]

Thus, the SC began work on the comprehensive assessment program assuming that the moratorium's main objective was to acquire more reliable cetacean research and management methods with the intention of considering a return to commercial whaling at some point.[66] However, the question of how much certainty was required or how little uncertainty could be tolerated before commercial hunts could recommence was never addressed by the commission in any detail, and its failure to do so has since caused considerable controversy over the IWC's purpose and objectives.

For several years after the moratorium's inception, and even during the negotiations prior to the vote, the means by which the moratorium was to contribute to more and better cetacean research remained unclear. Even the proposal for a comprehensive assessment of whale stocks was added as an apparent afterthought to the original moratorium proposal. Between the moratorium's adoption in 1982 and 1985, the SC repeatedly told the commission it did not understand what the comprehensive assessment was intended to mean and asked for an explanation and definition at each annual meeting. No answer from the commission was forthcoming. In 1985, therefore, the SC "decided that if progress was to be made it would have to define what it thought was a 'comprehensive assessment' and establish how it might be accomplished."[67] In April 1986, almost five years after the moratorium's adoption, a special meeting of the SC was held in Cambridge to develop a working definition of the comprehensive assessment and its aims. The IWC report of that meeting defines the comprehensive assessment as

> an in-depth evaluation of the status of the stocks in the light of management objectives and procedures. This could include examination of current stock size, recent population trends, carrying capacity and productivity. In order to achieve this the Committee agreed that it would need to:
>
> (a) review and revise assessment methods and stock identity; review data quality, availability requirements and stock identity;
>
> (b) plan and conduct the collection of new information to facilitate and improve assessments;
>
> (c) examine alternative management regimes.[68]

The ongoing debate over how the precautionary principle should manage scientific uncertainty, beyond simply banning all whaling indefinitely, is rooted primarily in its ambiguous nature and in the fundamentally different political positions of the pro- and anti-whaling protagonists. As a basis for policy, the precautionary principle is extremely general and provides little in the way of guidance in determining when and how much precaution is required.[69] The principle, as a result, became (and remains) an obstacle to the IWC moving beyond moratoria as a means of managing scientific uncertainty. The precautionary principle is simply too open to interpretation according to individual political interests and priorities.

Daniel Bodansky has noted, "Although the precautionary principle provides a general approach to environmental issues, it is too vague to serve as a regulatory standard because it does not specify how much caution should be taken."[70] This criticism is commonly made and it is consistent with the IWC experience. The two essential questions raised by the precautionary principle concerning (a) when it should be applied and (b) the level of precaution required and at what price are left mostly unanswered in the existing definitions quoted by policy makers, and the IWC has been no exception.

The task of defining "comprehensive assessment," a concept intended to reflect the reasoning behind the moratorium, was largely ignored by those supporting the moratorium; they seemed content with simply proscribing commercial whaling as a solution to the commission's management problems. A significant characteristic of the commission's resolutions and schedule amendments relating to the moratorium was the vague nature of the wording of key concepts and provisions. When comprehensive assessment was finally defined in 1986, the wording still left a great deal open to interpretation, in particular the description of comprehensive assessment as an evaluation of whale stocks "in the light of management objectives and procedures," which to this day remain far from clear.

By 1986, the IWC's management objectives were more confused than ever, as the commission by this stage was already totally divided over its interpretation of the IWC's parent treaty, the ICRW. Some governments that had supported the moratorium, such as Australia and New Zealand, were by now openly calling for a permanent end to commercial whaling on the basis of a recently developed nonconsumptive interpretation of "sustainable utilization." Meanwhile, those who had opposed it continued to insist that hunting should resume after the comprehensive assessment's completion. Thus, the environment in which the RMP was to be developed could offer no consensus on why it was needed or to what ends it would be used, if at all.

The two basic management objectives provided for by the ICRW are (a) that whale species are not hunted to extinction; and (b) that the maximum sustainable harvest is achieved. But as Donovan pointed out, these two objectives represent management extremes and require clarification in terms of how one is balanced against the other. Thus, the role of the SC in defining and conducting the comprehensive assessment was to provide the commission with specific options based on various "trade-offs" between these two basic objectives.[71]

Donovan, however, also described the "setting of objectives and the relative weight given to those objectives (the trade-offs)" as a process that requires "political rather than scientific decisions."[72] By making this distinction, which is often made in explanations of why the RMP is yet to be implemented, Donovan implied that the process of formulating the various options and trade-offs is scientific and does not become political until the commission decides which combination of options should be given priority. But the basic objectives of the ICRW, which set the parameters for the kinds of advice offered by the SC (i.e., the undesirability of extinction and the desirability of sustainable utilization), are essentially political judgments that, like the precautionary principle, must influence the SC's research into providing options for managing scientific uncertainty.

The development of the RMP was largely determined by the precautionary principle and the ways in which governments opposed to commercial whaling chose to interpret it—primarily because the level of certainty required by policy makers generally is inversely related to the level of political support available for the policy in question. Put another way, the less political recognition research findings have in terms of (a) fulfilling a perceived need (criterion I) and (b) not conflicting with other needs (criterion II), the greater the likelihood of the precautionary principle being invoked in opposition to the recommendations of that research.

In extreme cases such as in the late 1950s and early 1960s, when rapidly declining pelagic catches convinced whaling governments that many stocks were nearing extinction, the SC's advice to reduce catch limits was eventually accepted because of broad political acceptance of a recognized certainty—rather than uncertainties. But even in such cases, political agreement on the issue (i.e., species should not be driven to extinction) was still required before scientific advice was able to directly influence the commission's policies. No political consensus on the desirability of the RMP as a means of actually managing commercial hunting existed at any time during its development and, despite its belated adoption by the commission in 1994, anti-whaling governments continue to oppose its implementation.

THE RMP: AN OVERVIEW

The comprehensive assessment pertained directly to the future effects of the moratorium on stock numbers and was essentially an attempt by the IWC to start over in terms of its management of whale stocks. The assessment was the starting point for the development of a new management regime, intended to incorporate the lessons learned from the NMP and the conservative approach toward uncertainty set out by the precautionary principle. The various parts of the new regime exist under the umbrella of the Revised Management Scheme (RMS)—which provides the yet to be agreed upon monitoring and policing procedures that the RMP would operate under[73]—and include the following: (a) the comprehensive assessment, which provides data on stock numbers, carrying capacity, productivity, and population trends; (b) the RMP, which consists of the catch limit algorithm (CLA: the methodology for calculating catch limits) and the various rules determining how the CLA is applied (i.e., when reviews on information should be carried out, details of stock boundaries, what to do if too many whales of one sex are caught); and (c) international observer and monitoring schemes—yet to be completed—intended to monitor catches made under the RMP and to ensure that products from illegal catches are not made commercially available.

The comprehensive assessment's main goals of providing a detailed evaluation of whale stocks and developing a replacement management regime for the NMP (the RMS) were closely linked by the need for the limitations of one (stock assessment) to be accounted for by the other (the RMP). The interdependency between stock estimates and the RMP is well illustrated by Donovan's description of the perfect procedure as "one which arrived at safe catch limits and required no information!"[74] Given the SC's use of estimates based on 95 percent confidence intervals (i.e., an interval calculated from the sample that contains the true population parameter in 95 percent of all samples), which, depending on the size of the stock, could spread over tens of thousands of animals (or in the case of the minke, hundreds of thousands), the RMP needed to be able to work reliably with a minimum of information.

The stock estimates used in the RMP are weighted to address the unavoidable uncertainties in whale population estimates, and are based on data from sighting surveys at regular intervals, any recent catches if taken, and catch history records. Unlike the NMP and most fisheries management programs, the RMP does not use models based on catch per unit effort or biological parameters (natural mortality rates, pregnancy rates, and recruitment rates) to simulate the behavior of exploited stocks. The use of CPUE assumes a direct relationship between CPUE and population size but is made problematic by a number of variables, including the efficiency of different fishing efforts, the influence of weather conditions, and also its past failure to show a statistically significant decline in stocks before they had already become severely depleted.

Biological parameters have also been unreliable guides to population dynamics due to problems in collecting data and the inability of scientists to obtain findings that met the level of precision required by the NMP. As a result of these problems, CPUE and biological parameters were largely discounted as methods for calculating relative abundance during the comprehensive assessment.[75] Thus, it was essential for the RMP to be able to account for high levels of uncertainty with a low level of risk—a feature that reflected the commission's newly acquired preference for precautionary management principles.

The RMP is based on many of the same principles that were introduced by the NMP, although at the time of writing it is intended for the management of baleen whales only. Management by individual stocks and species (rather than collectively as with the Blue Whale Unit), the setting of catch limits on the basis of carrying capacity and optimal yield, and the imperative that stocks should be protected before they become endangered are all principles that were first introduced by the NMP and are included in the RMP.

The RMP's most distinctive feature is its more conservative management of uncertainty due to the CLA's basic assumption that the stock information upon which its calculations are based is flawed. But while the IWC's precautionary approach to management had become far more conservative and risk averse than ever before, it still formally embraced the fundamental assumption that commercial whaling was acceptable—a position reflected in the three main objectives the commission assigned in 1981 for future management policies:

(1) to ensure that the risks of extinction to individual stocks are not seriously increased by exploitation;

(2) to maintain the status of whale stocks so as to make possible the highest continuing yield so far as the environment permits; . . . never to move individual harvested stocks or groups of stocks of the same species in a direction which reduces its or their combined sustainable yield;

(3) to ensure the maintenance and orderly development of the whaling industry.[76]

These objectives later became the basis of the goals the commission would assign to the RMP, which Donovan has summarized as follows:

(1) Catch limits should be as stable as possible;

(2) Catches should not be allowed on stocks below 54 percent of the estimated carrying capacity (as in the NMP) [with 54 percent being the lower limit, set as a safety feature, of the estimated 60 percent MSY level for an unexploited stock];

(3) The highest possible continuing yield should be obtained from the stock.[77]

Of these three objectives, the commission placed the greatest emphasis upon the second—indicating that the IWC's preferences for stable and continuous yields should always be secondary to its preference for no stock to ever be depleted below its optimal reproduction level.

The CLA component of the RMP is based on a feedback mechanism controlling the data it uses for catch limit calculations. At first, the best available estimates (from previous catch records, regular sighting surveys, and catches if available) are used and then improved upon as further information becomes available. This particular procedure, developed by British scientist Justin Cooke, was summarized by the Comprehensive Assessment Workshop on Management in 1987 as being

> based on using a provisional value for the MSY exploitation rate ... along with a decision rule specifying what proportion of this rate can be taken, according to the estimated degree of depletion of the stock. This exploitation rate is converted to a catch limit using an estimate of current population size. The estimate of current population size is obtained by averaging survey estimates of the population obtained over the most recent 10 years, after adjusting each annual estimate by the amount of subsequent catches. This filtered estimate is then adjusted for uncertainty by taking its lower 95 percent confidence limit. The depletion of stock is estimated by fitting a population model through the estimate of current population size using the known catch history and the provisional value for the MSY exploitation rate.[78]

The feedback mechanism of Cooke's procedure is an important feature because it allows for the use of inaccurate information (i.e., stock numbers past and present) by setting very conservative parameters (MSY rates and levels), which are initially assumed but only permit very conservative catch limits. As more stock data from surveys and catches become available, the MSY rate and level for the stock is adjusted accordingly. The procedure is, in effect, self correcting and claims no more than a 5 percent chance that catches would be taken from a stock more than 10 percent below its most productive level, that is 54 percent of the estimated preexploitation population—the lowest population number at which MSY can be safely achieved. Stocks below 54 percent of their original population are classified as Protection Stocks and no catches are allowed.[79]

The MSY rate (MSYR) is a key factor in determining catch limits. According to IWC Scientific Committee member Doug Butterworth, the MSY rate "is the ratio of the annual maximum sustainable yield to the population size at which that yield is obtained (conventionally taken to be 0.6K for baleen whales). Evaluations of risks associated with the RMP have all been based on an MSY rate of 0.66 percent. However, the appropriateness of this choice is coming under increasing question, as all recent evidence points towards baleen whale stocks having MSY rates of about 3 percent or more."[80] Thus, at 0.6K (i.e., 60 percent of the estimated original population), the MSY for a given stock (i.e., the number of whales that can be taken at 0.6K without depleting the stock) is assumed to be 0.66 percent of whatever population figure 0.6K represents (i.e., MSY = MSYR x 0.6K).

Cooke's procedure was chosen by the commission out of a choice of five on the recommendation of the SC, after a series of computer trials tested its ability to allow for large errors in population estimates and uncertainty over carrying capacities, environmental damage, and recruitment. The CLA also incorporates two important assumptions into its calculations: (a) population estimates can only provide a range within which the true population size probably exists (i.e., the 95 percent confidence interval mentioned earlier); and (b) all estimates are likely to be biased. The IWC's whale population estimates, used as input for the RMP, are based on guidelines for sighting surveys and analyses of data approved by the SC.[81] Estimates are drawn from within a 95 percent confidence interval incorporating a measure of uncertainty of the estimate expressed as a coefficient of variation (CV), which has an inverse relationship with the possible catch limit. In other words, if the CV is high it indicates a higher level of uncertainty connected with an estimate and therefore a lower possible catch limit. According to Holt,

> All catch limits depend ... on a measure of the uncertainty of the estimates of current numbers, expressed as a coefficient of variation (CV). The higher the CV the more uncertain the estimate and the smaller will be the catch limit, and vice versa.
> The CV of a single survey estimate is called the *observation error* [CV(o)]. In surveys carried out as part of the International Decade of Cetacean Research (IDCR) in the Antarctic CV(o) is typically about 0.25. This means that the 95 percent confidence limits for the estimate roughly embrace plus and minus 50 percent of the "best" estimate. There is (only) one chance in twenty that the true number will lie outside that confidence range. The CV(o) is, in IDCR cruises, rarely less than 0.2 but in 10 out of 27 half-Area surveys it has been higher than 0.3.[82]

On the basis of the available best estimates of the three types of data the CLA requires—abundance estimates, the cumulative catch history and its distribution over time, and survey variance estimates (the CV)—it will calculate a catch limit for the stock that is adjusted as new data become available. This catch limit is intended, over time, to maintain the stock at (or allow it to reach) a level within the 72 to 54 percent range of estimated original population originally set out under the NMP, depending on the "tuning level" from within the range chosen by the commission. The tuning level is a benchmark used to set the desired level at which the stock will stabilize. The tuning level for the RMP is set at 72 percent (a higher tuning level results in lower catches), which means that after one hundred years of catching, the stock level should be at or around 72 percent of its estimated original population.[83] The IWC's choice of a 72 percent tuning level reflects the precautionary climate within the commission and the majority preference for risks to stocks to be minimized as far as possible. As Cooke has noted,

> The RMP differs from the previous attempts to manage whale stocks in several ways, one of which is that it only makes use of data which we know are obtainable, and secondly it provides specific rules for determining what levels of catch are safe based on these data. So once the data are available, there's little need to discuss what level of catch would be safe.[84]

The RMP does incorporate important elements of the NMP and is also based on the concept of MSY (i.e., that stocks have higher rates of reproduction at some point below the maximum population level afforded by the environmental carrying capacity).[85] But the RMP is at the same time very different from the NMP in that it specifically addresses the uncertainties that are inherent to the data upon which it relies. According to Donovan,

> The Revised Management Procedure is essentially a safe way to calculate catches. And the reason we can say it's safe is because we've used computer whales . . . [to] simulate all the different kinds of scenarios that we can imagine . . . to do with almost anything—abundance estimates wrong, number of catches wrong, environmental conditions changing dramatically, disasters happening so the population falls by half in a single year—and we run the model and see if our way of calculating catch limits is safe and will cope. And effectively we've run millions of simulations, and I personally

think that the Revised Management Procedure is by far the best tested way of managing any natural resource.[86]

The remaining component of the RMP deals primarily with requirements for how sighting surveys should be conducted and the data interpreted in addition to the important issue of how stock boundaries should be determined. The CLA is designed to be applied to isolated stocks of whales and, therefore, the RMP has to deal with any uncertainty involving the placement of stock boundaries, since errors could lead to severe stock depletion. Some earlier methods used under the NMP to estimate current and initial populations relied on biological parameters, which assumed that management stocks were the same as biological stocks.

The issue of stock identity became a priority for the SC during the comprehensive assessment, since completing the RMP required a resolution. The solution was to set much smaller catch limit areas than had been used under the NMP, based on the known or suspected biological information, before then testing the effects of potential errors in stock identity with simulations. At the time of writing, work on this problem, known as "multistock rules," has only been done for minke whale stocks in the North Atlantic and Southern Ocean (which currently limits the RMP's potential application to only these stocks). This has resulted in the NMP's 60 degrees east longitude catch limit sectors being reduced to catch limit areas for each 10 degrees east sector.[87]

In short, the RMP was designed to balance conservation and exploitation by determining conservative catch limits requiring only a minimum amount of data. By recognizing the shortcomings and inaccuracies of available stock estimates, the CLA factors in a high degree of scientific uncertainty in order to allow endangered stocks to regenerate to 72 percent of their original level (calculated on the basis of catch histories and current population estimates) while allowing no stocks to fall below 54 percent of that level, in accordance with the MSY classifications set out under the NMP.

By the early 1980s, recognizing and dealing with scientific uncertainty was an omnipresent policy consideration within the IWC. This approach to management resulted from the commission's general acceptance of the precautionary principle in its endorsement of policy objectives. The political environment of the IWC had indeed changed since the UN conference in Stockholm, and the extent and direction of this change was reflected in both the policy objectives of the commission (i.e., conservation before exploitation and the need to manage uncertainty) and the kind of scientific research and advice that these new objectives were now producing (i.e., the RMP).

GREEN POLITICS AND SCIENTIFIC UNCERTAINTY

The arrival of the moratorium, with its scheduled implementation for the 1985–86 season, meant that all commercially hunted stocks would be protected for the first time. The political climate of the IWC was such that the overwhelming majority of its members would no longer accept the level of scientific uncertainty that had so far characterized cetacean research (and fisheries research in general). The growing political pressure for less uncertainty—generated by the new criterion I and II priorities adopted by the bulk of the commission's membership—led to a paradigm shift in terms of how the SC formulates its advice for the commission; a shift manifested in the precautionary principle requirement that the management of commercial hunting must demonstrate it will not pose any significant threat of depletion to stock numbers.

According to John Lemons, scientific method traditionally has preferred minimizing acceptance of Type I errors (false positive results, e.g., mistaken predictions that an activity or substance is harmful) over acceptance of Type II errors (false negative results, e.g., mistaken predictions that an activity or substance is harmless) when evaluating a testable hypothesis. This is because Type I errors, according to accepted norms, are assumed to lead to "speculative thinking" and are also believed to contain a higher likelihood of error.[88] In other words, it is possible to demonstrate through inductive reasoning that something is harmful, but it is impossible to demonstrate that something is harmless, due to the finite nature of experience.[89] Thus, if a given activity or substance cannot be shown to be harmful, the traditional view has attributed a lower likelihood of error to the conclusion that it is, therefore, harmless (Type II) than to the alternative conclusion that it may nevertheless be harmful (Type I) even though we cannot demonstrate it.

In contrast, the precautionary principle—by shifting the burden of proof and requiring those supporting an activity to demonstrate it will not cause harm—has forced the SC to conduct and test its research in a way that minimizes Type II errors by accounting for a myriad of potentially harmful, but less likely, Type I effects. This approach, resulting from the altered political priorities of cetacean management in particular and in fisheries more generally, has fundamentally changed the nature of cetacean research by turning the role of scientific uncertainty in policy making on its head—a development illustrated by the objectives and standards the commission set for the RMP's development and its eventual adoption.

Given the wide agreement among scientists that the RMP represents a major advance in wildlife management and that it is also of great potential benefit for fisheries management in general, the "positive" effects of the precautionary principle's application in the case of IWC management should not be ignored. Indeed, an increasing number of biologists and other scientists share the opinion that the precautionary principle has some role to play in future environmental management, given the "pervasive scientific uncertainty surrounding the understanding of biodiversity, and science's emphasis on minimizing type I error."[90] According to Lemons,

> The recommendation to adopt . . . a precautionary principle and shift the burden of proof to those undertaking activities that might cause harm to biodiversity to *prove* that the activities will not cause harm is consistent with views that it is better to minimize type II error in conservation decisions. In other words, from an ethical perspective, it is more prudent to accept a higher risk of an erroneous conclusion that activities will cause harm than it is to accept a lower risk of a false null hypothesis that no harm will result from activities that potentially threaten biodiversity.[91]

As mentioned earlier, however, application of the precautionary principle to policy decisions is complicated by the principle's lack of clarity and also its ability to be defined by any number of judgments concerning the status of cetaceans and their relationship to humans. As Donovan has observed, the precautionary principle's lack of clarity on important issues, such as acceptable levels of risk, allows policy makers and politicians to be selective about when and how it is applied, a situation that often leads to instances of inconsistency in environmental policy and accusations that the precautionary principle is "antiscience":

> In short, the precautionary principle is a good basis for policy formulation when managing natural resources. What is important is how it is applied. Management is essentially an exercise in trade-off—balancing the utilisation of the resource with the risk of its depletion. The difficulties lie in determining what is an appropriate balance. This is essentially a polit-

ical decision which must be taken in the light of scientific advice. This was the case in the RMP where the Committee provided the Commission with three possible "tuning" values.... Determining an appropriate level of acceptable risk allows politicians a "free-rein" and this is where outside factors (e.g. being popular) can conflict with issues of consistency with the management of other natural resources (e.g. fisheries). I believe that the work of the Scientific Committee in explicitly taking uncertainty into account in management is exemplary, even though it can be argued that the ultimate "tuning" level chosen may be extreme.[92]

As the precautionary principle is aimed squarely at increasing environmental protection, and is also open to various interpretations concerning when and how it should be applied, it is no surprise that the principle's ascendancy paralleled that of the many environmental NGOs that have appeared since the early 1970s. And given the management problems facing the IWC by the end of the 1970s—a history of severe overhunting, disagreement within the SC over the reliability of data and the NMP's inability to account for uncertainties—it is also not surprising that the precautionary principle became such an effective weapon in countering the SC's preference for minimizing Type I error when formulating its advice to the commission (i.e., that no scientific evidence existed to support the adoption of the moratorium).

By 1979, the year the IWC endorsed the Seychelles' Indian Ocean Sanctuary proposal, many IWC members had made their distrust of the SC's advice quite clear. In effect, the shortcomings of the science upon which the NMP was based were undermining the arguments that had so far prevented the moratorium from being adopted. Governments opposed to the moratorium were being forced to directly confront the precautionary principle and its emphasis on scientific uncertainty for the first time.

In his opening statement to the 1979 IWC meeting, U.S. commissioner Richard Frank clearly illustrated the growing suspicion of the NMP by openly criticizing the procedure and the data it used:

> The deficiencies in the current management system are manifold. For example, when there has been a history of harvesting at a constant level, the system allows continued harvesting at that level, even though actual MSY data is scant, thereby creating a risk that an otherwise healthy stock will be reduced to protection status.[93]

The World Wildlife Fund (WWF, also known as the World Wide Fund for Nature) had been expressing concern over the IWC's management of cetaceans since the late 1960s and was one of the first environmental lobby groups to provide detailed criticism of the NMP. Other NGOs opposing whaling, including Greenpeace and the Fund for Animals, had also joined the WWF in its criticism of the SC's advice to the IWC by the mid-1970s. But these groups had expanded their arguments against the commission's policies to include moral and ethical concerns and the support of growing public opinion. Some IWC members, most conspicuously Australia and New Zealand, were quick to endorse the growing feeling in many Western countries that all hunting of whales, in addition to being based on unreliable scientific research, was also inhumane and ethically unacceptable.[94]

Thus, by the late 1970s, both commercial and aboriginal subsistence whaling was being opposed on two fronts, and public opinion had become a significant factor in the moratorium debate. And while the moratorium on commercial hunting was finally adopted only a few years later in 1982—ostensibly because of scientific uncertainty—the growing public perception of cetaceans as a generic environmental symbol had already been established by environmental NGOs, to the point where it would soon be used by some IWC members to call for a permanent ban on whaling. In spite of its shortcomings, the precautionary principle indeed had been effective in prompting the creation of the RMP and also in helping to bring a temporary end to commercial whaling. But its role in determining the moratorium's future would, in effect, soon be overshadowed by an issue that was fast eroding the relevance of scientific uncertainty in the shaping of IWC policy: the anthropomorphism of cetaceans.

ENTER THE SUPERWHALE

Public perceptions of whales have changed significantly since the early 1970s, with the most dramatic changes occurring in the IWC's former whaling member states. Indeed, the impact of this change of attitudes on IWC policies over the last three decades should not be underestimated, particularly when examining the extent to which the commission's (ongoing) aversion to commercial whaling has been based on issues of scientific uncertainty.

On the surface it appears that for much of the last three decades, the

IWC has been split between two distinct blocks: the pro- and anti-whaling members. However, seeing the IWC protagonists as either conservationist or preservationist portrays a much more complex political environment within the IWC and reveals a clearer picture of its cast of characters. The conservationist parties within the IWC are generally those subscribing to the idea of "wise resource use" and have no problem with the goal of sustainably exploiting natural resources or wildlife for either consumptive or nonconsumptive purposes (i.e., whale watching). For conservationists, the main objective is to avoid overexploitation of a resource and the subsequent extinction of any species. IWC members or participating NGOs in this category can be regarded as any government or group that supports some form of whaling, either in practice or principle, providing it is conducted in a sustainable manner with an acceptably low level of risk of endangering the exploited species.

Preservationists, on the other hand, argue that the treatment of whales as an exploitable resource is unacceptable (although most would condone whale watching). Most preservationists can be identified as either supporters of animal rights or people who will accept virtually no risk of a species becoming depleted as a result of lethal exploitation. Animal rights–orientated preservationists, however, do not necessarily extend their concerns for protecting animals to all species. While there are preservationists who oppose the killing of any species, whale protectionist advocates most often "take their rhetoric from the animal rights philosophers and explicitly limit it to cetaceans. . . . To these people, whales are uniquely special, based upon claims of their biological, ecological, cultural, political, and symbolic specialness."[95] Thus, in the context of the IWC, the policies of member governments and related NGOs proscribing any return to commercial whaling are essentially preservationist and are based on the view that because whales are special, they require different treatment to other animals.

Paul Spong, a New Zealand psychologist based in Canada, is often credited as among the first to publicize the view that cetaceans are highly intelligent and, therefore, should be protected from hunting. Spong was a "key figure" in turning Greenpeace's attention to whaling in 1974, following a change in leadership within the organization that led it from a primarily antinuclear focus to "the plight of the great whales."[96] According to Spong,

> Whales are icons and symbols that have huge potential for arousing public interest, concern, and sympathy for the entire range of environmental

issues. That's why the first UN conference on the environment, in Stockholm in 1972, chose the whale as a symbol . . . of both crisis and hope. . . . I certainly agree that there is no need to kill whales, period. Whales are far too interesting as live animals for us to enjoy and try to understand for us to demean ourselves by abusing them. The moral issue is simple, not complex. Whales are the highest evolved forms of life in the ocean. We are only beginning to glimpse their magnificence, their intelligence, and the crucial roles they play in the ocean; but we do know one thing for sure . . . they deserve our respect and protection. It is their due.[97]

It was based on such observations concerning the nature of cetaceans that the concept of what pro-whaling critics have termed the "superwhale" materialized. The late 1970s saw the popular image of whales transformed by some environmental groups, most notably Greenpeace, in order to generate public and political support for the moratorium and to add further weight to the scientific uncertainty arguments against whaling within the IWC. Ben White, international coordinator for the Animal Welfare Institute, described the importance of whales to the environmental movement and their relationship with humans in the following terms:

Whales have become our environmental symbol for saving the world. . . . Those of us who have seen whales eye to eye have found or believe that they are persons, and therefore we have a greater family of life on earth; that we are not alone.[98]

According to Arne Kalland, a social anthropologist at the Nordic Institute of Asian Studies in Copenhagen, the superwhale is a composite of various species and is "endowed with all the qualities we would like to see in our fellow humans: kindness, caring, playfulness."[99]

Whale protectionists tend to talk about the whale in the singular, thereby masking the great variety in size, behavior and abundance that exists among the 75 or more species of cetaceans. By lumping together traits found in a number of species an image of a superwhale is created. Thus "the whale" is the largest animal on earth (this applies to the blue whale), it has the largest brain on earth (the sperm whale), a large brain-to-body-weight ratio (the bottlenose dolphin), it sings nicely (the humpback whale), it has nurseries (some dolphins), it is friendly (the gray whale), it is endangered (the blue and right whales) and so on.[100]

In Kalland's view, the superwhale is both an effective marketing symbol for NGOs opposed to whaling and an effective means for lowering public tolerance for scientific uncertainty and research supporting the notion of sustainable hunting.[101] He argues that by promoting the superwhale image, environmental groups have redefined whales as a highly marketable "green" commodity from which they are able to gain considerable economic benefit through attracting additional members and selling "environmentally friendly" seals of approval to corporations wanting a greener image: "The whaling issue has become a symbol to the environmental and animal welfare movements because this issue provides them with an easily identifiable enemy and a sense of urgency, two factors a consultant to Greenpeace identified as the requirements for raising money."[102]

Others, including environmental organization leaders and IWC scientists, have made similar observations and cited numerous examples of this particular marketing strategy.[103] Heidi Sorensen, the leader of Friends of the Earth affiliate Nature and Youth, has noted that several international environmental groups "have spun the myth of the Super Whale" in order to better market their organizations.[104] She gives the following example of how the superwhale has been used as a money-making commodity:

> In October last year [1993, the year Norway unilaterally recommenced minke hunts in the North Atlantic], Greenpeace Austria placed a large obituary in the major newspapers saying "It is with great sorrow we announce that our friend and brother the MINKE WHALE, has passed away after being brutally murdered. Instead of wreaths and flowers, please send money to the environmental organisation Greenpeace's Whale Rights Campaign."[105]

In terms of its influence on public perceptions of scientific uncertainty and risk in relation to commercial hunting, the superwhale concept has played a major role in questioning the reliability of scientific research aimed at providing sustainable hunting methods. By elevating whales above other creatures, and also by providing the impression that many species are on the edge of extinction, environmental groups and likeminded governments have effectively raised the crossbar of the criterion II priority (that requires research to avoid compromising established needs). Research and industry objectives that are not compatible with the recently established desire in Western societies for only nonlethal exploitation of cetaceans are, according to many scientists, increasingly opposed for reasons that have little to do with conservation. Gulland, for example, noted that,

> Whales ... make excellent fund raisers, probably behind only giant pandas and baby seals. There may no longer be urgent reasons of conservation for continued pressure to strengthen the controls on whaling, but there are sound financial reasons for groups that depend on public subscriptions to be seen to be active in "saving the whale."[106]

The popularity and importance of whales within Western societies, as both a symbol for the fight against environmental destruction and also as a commodity to be enjoyed in unprecedented ways (i.e., whale watching, whale books, whale movies, whale songs), grew rapidly in conjunction with the further development of the superwhale concept during the 1980s. So much so that by the early 1990s, when the RMP and some elements of the comprehensive assessment were nearing completion, the status of "the whale" was such that any research aimed at sustainable hunting was by now anathema in the eyes of the many governments and environmental NGOs. By this stage, these anti-whaling elements had assumed an entirely protectionist position (as opposed to conservationist as discussed earlier), based on the claim that this stance reflected public opinion. Thus, by the early 1990s, concerns about scientific uncertainty, which had played such an important role in the moratorium's adoption and justification, were being expanded by anti-whaling governments and NGOs to include ethical arguments concerning the lack of humaneness of killing methods and the immorality of killing allegedly highly sentient creatures such as cetaceans.

By definition, the protectionist position makes irrelevant any scientific evidence in support of sustainable hunting, which in effect means that even if uncertainty could be reduced to zero, something science cannot do, protectionists still would not accept any return to hunting. The engine driving this argument is the perception of whales as sentient, intelligent creatures that are different from other animals and also their status as battle flag in the war against environmental degradation—an engine, metaphorically speaking, that runs on the highly combustible fuel produced by the superwhale concept.

So, on the one hand, the idea of the superwhale has proven an effective tool in downgrading the relevance of scientific research intended to create a sustainable whaling regime, regardless of the level of uncertainty involved (i.e., whales are special, therefore it is wrong to hunt them). On the other hand, the precautionary principle continues to be invoked by protectionist-orientated opponents as a means of questioning such research, to further ensure commercial whaling does not recommence. The duality of this approach allows protectionist proponents—most notably Australia, New Zealand, the United Kingdom, and France, in addition to NGOs such as Greenpeace, the International Fund for Animal Welfare (IFAW), and the WWF—to keep one foot in each the protectionist and conservationist camp in the course of unconditionally opposing whaling. The most likely explanation for this strategy is a reluctance on the part of these governments and groups to divorce themselves entirely from the credibility that "science-based" arguments can offer, even though many of the uncertainty arguments used to question the RMP also can be seen as attempts to erode confidence in science as part of the solution to environmental problems.

Both Kalland and Butterworth, among others, have accused critics of the RMP and the SC's abundance estimates of duplicity in their use of scientific uncertainty.[107] Kalland, for example, points out that science-based ecological arguments are "more palatable for various reasons than are ethical and moral ones to a number of people, corporations, and government agencies." Furthermore,

> many protectionists are more than reluctant to change their rhetoric from an ecological discourse to animal welfare or animal rights arguments. Thus the myth of the endangered whale is sustained by charging scientists producing new and larger stock estimates of being incompetent, biased and "bought" by governments of whaling nations, by refusing to accept these new population estimates or refuting their relevance, or by introducing new arguments into the ecological discourse.[108]

Clearly the uncertainty arguments made against research supporting a limited return to whaling under the RMP inspire little confidence in scientists and their ability to gather and correctly interpret empirical data. It is important to note, however, that both sides have used the mantle of scientific credibility in order to justify their actions. In the course of making the precautionary principle applicable to the RMP, scientific argument based on alternative analyses of the same data has been a useful tool in persuading policy makers and the public that sufficient uncertainty does indeed exist to justify postponing its implementation, particularly when the onus of proof that overhunting will not occur lies with those who support the RMP's use. Opponents of whaling also have used various research efforts to support their claims about the specialness of whales (i.e., their intelligence, humanlike behavior, etc.), thereby lending scientific credence to the idea of the superwhale, the moral arguments against whaling, and also to the perception that little or no uncertainty is acceptable and that, therefore, the moratorium should remain in place.

Supporters of the RMP and its implementation argue that their interpretation of the data is correct but of course have no way of proving it to their opponents, not least because of the inability of scientists to conclusively prove any theory or empirical method true, but primarily because any and all attempts to do so simply are not compatible with the protectionists' criterion I and II priorities. In an attempt to counter the popularity of the superwhale concept, pro-whaling governments (in particular Japan) and NGOs have responded by promoting the cultural importance of whaling to their societies. This approach, however, has been largely unsuccessful. It has made the pro-whaling lobby dependent on the strength of their scientific arguments in order to force change. But such efforts are hamstrung by both the inability to remove uncertainty (concerning Type I and II errors) and in particular the success of the superwhale concept in making science and uncertainty mostly irrelevant to the protectionist position on commercial whaling.

The reliance of the whaling governments on science as the basis of their arguments, however, has been useful in justifying limited hunting after the moratorium took effect under the ICRW provision that allowed governments to issue permits for scientific whaling. The practice of taking whales for scientific research has become one of the major controversies within the IWC and, with the exception of the debate over the RMP and its implementation, represents the clearest manifestation of how the commission's polarized political environment has made an issue of what is or is not valid science. In the view of Japan—the most enthusiastic IWC member in issuing scientific permits to its nationals and in conducting both lethal and nonlethal research—the research it conducts is scientifically valid and completely legal. But in the view of the anti-whaling governments and NGOs, Japan is simply using science to exploit a "loophole" in the moratorium and continue whaling.

Again, a key factor in explaining the ongoing conflict over whaling is scientific uncertainty and its potential as a double-edged sword in debates concerning the reliability of scientific advice and data. The protectionist strategy of employing the precautionary principle to demand more proof that whale stocks are robust enough to justify the RMP's implementation necessarily implies the need for more research, thereby providing the Japanese with some justification for issuing scientific permits. The problem, however, is not simply that more information is needed, but rather what kind of research is needed and what kinds of questions need to be answered. Scientific method alone cannot answer these questions, since value judgments concerning what is desirable are required to define the policy objectives scientists need in order to prioritize their research goals.

It is, perhaps, more than coincidental that the ambiguity and controversy that has surrounded Japanese scientific whaling is similar to that which characterized the goals of the moratorium and the lack of direction provided by the commission for the comprehensive assessment. In the absence of clearly defined policy objectives, scientists can hardly be expected to provide relevant advice; the inability of the IWC to provide this direction may go some way toward explaining the increasing irrelevance of the SC's work. As Butterworth has observed,

> The real debate in the IWC has been between some countries wishing to preserve industries, employment and a food source based on whales, and others wanting these animals classed as sacrosanct. The terms of the convention [the ICRW] have required that this debate be conducted in a scientific guise, so that these hidden agendas have had to be played out in the scientific committee.[109]

The role of science in the IWC and the interpretation of uncertainty has been left in a form of limbo since the early 1990s, when ethical and moral issues were first introduced to seriously question the usefulness and appropriateness of the research undertaken during (and after) the comprehensive assessment. As debate over the future of the RMS continues, scientific advice appears to be both relevant and irrelevant to varying degrees, depending on the argument being made and the objective in question. Indeed, the extent to which scientific advice, rather than the various political objectives of the commission's members, has influenced the IWC's policies is very much open to doubt, as are the equally important questions of how science is used and to what ends is it used.

6

CONCLUSION

The end of chapter 1 outlined a hypothetical situation concerning the discovery of minke whale oil as the source of an effective and safe cancer treatment and asked how such a discovery might affect the debate over uncertainty issues concerning population estimates, MSY rates and levels, and so on. The answer, I believe, is now clear. Such an eventuality undoubtedly would cause a major change in the treatment of uncertainty by all the governments currently opposing any lifting of the moratorium: the utility of exploiting minke stocks, in criterion I terms, in order to provide a cancer cure would far outweigh the existing benefits derived from not exploiting them.

In fact, the political advantages many governments derive from protecting the minke would disappear as the criterion II definitions of established needs in relation to whales abruptly changed. Few but the most devout animal protectionists would dare to openly argue in favor of protecting an unendangered species at the expense of human lives. Subsequently, the IWC's ongoing disputes over uncertainty in population estimates, effects of environmental change, MSY levels, and monitoring of catches would disappear almost over night as governments quickly readjusted their perceptions of uncertainty to accommodate the needs of their societies.

As this hypothetical scenario demonstrates, the strength of the data and the conclusions made by scientists generally concern politicians less than the political impact that acceptance or rejection of a given piece of research is likely to create. This perspective, when applied to environmental policy debates like those of the IWC, can explain a great deal about why perceptions of how to manage uncertainty vary widely according to the issues at stake. As Aron has observed concerning the strong opposition to the NMP and also the RMP, it is the political implications of the issue at hand rather than the science behind it that often have the most influence on the positions governments take:

> The NMP, unlike the RMP, did not explicitly take errors into account in its formulation, however, in generating critical numbers, the low end of the estimates was selected to minimize the consequences of error. Also the terms of the NMP were, by themselves, very conservative. The big problem was less the data issue, which was solvable, but more the fact that the NMP permitted whaling. This is the real issue of the RMP, as well.[1]

As the preceding chapters have shown, the concept of scientific uncertainty, including the risks it poses and judgments concerning where the burden of proof should lie, has been understood and applied in different ways and toward different ends over time in the IWC. The IWC's first two decades were characterized by policies that, for the most part, used uncertainty as a basis for downplaying the risk of overhunting, resulting in the severe depletion of most Antarctic stocks. An increasing number of governments in the following decade, however, adopted a very different view of uncertainty and demanded progressively smaller quotas before finally opting for no commercial hunting at all.

For many IWC members in the 1950s and early 1960s, dealing with uncertainty issues was a simple and straightforward task. For the five Antarctic whaling nations, uncertainty concerning population estimates and quota levels appeared to mean only that there were no compelling reasons to adopt lower quotas at the expense of their respective industries. This approach to uncertainty dominated IWC policy until the mid-1960s when two important changes occurred. The first was a consensus among the Antarctic nations on the need to reduce the Antarctic quota, brought about only by the increasing scarcity of the larger species. The Scientific Committee's improved ability, resulting from the work of the Committee of Three and Committee of Four, to better explain the cause of the decline in whale numbers and catches (i.e., unsustainable quotas) also was an important factor here, but would in itself have made little difference without the added weight of the Antarctic collapse. The second change, in part a result of the first, was the declining financial viability and importance of Antarctic whaling for all but the Japanese and Soviet industries, which reduced not only the number of active Antarctic whaling nations but also the strong influence that the whaling industry had so far exerted over commission policy.

The impact of these changes was such that, by the early 1970s, many in the IWC either had or were about to adopt an entirely different perspective on what uncertainty should mean and how it should be represented in commission policy. The emerging view within the commission following the UN call for a ten-year moratorium in 1972 was that stocks, rather than the remaining industries, should be given the benefit of any doubt in the setting of quotas. This landmark change in perception was further encouraged by the increasing political influence of various environmental groups, some of which proved extremely effective in mobilizing public opinion in support of the UN proposal and in bringing external pressure to bear on both the commission and its SC.

The most significant change within the IWC during the 1960s was that the majority of commissioners appeared to be far more enthusiastic in their support of the majority scientific advice on offer. But, as noted in chapter 4, this change in attitudes was very slow in coming. The shift should in any case be attributed more to the sharp decline in the viability of whaling during the period than to governments suddenly accepting the SC's advice and privileging the long-term interests of the resource and industry above the short-term needs of their respective whaling industries.

By the early 1970s, the criterion I priorities of many whaling nations had dramatically changed in tandem with the fast-decreasing relevance of whaling to their perceived national interests, which hitherto had been to catch as many whales as possible. Whereas government attitudes toward scientific advice and the uncertainties it entailed had been determined by perceived national interests, attitudes within the IWC came to be determined by the decline of whaling and its replacement by a new set of benefits—those that could be derived from exploiting whales as a political, rather than natural, resource. Indeed, this dramatic shift in how whales and whaling were perceived represented the real sea change in the IWC: an about-face on the part of several former whaling nations that ushered in the adoption of the precautionary principle approach to management, promising dividends that would become manifest over the next decade in the minds of all but the IWC's dwindling minority of whaling advocates.

With the IWC's implementation of the NMP in 1975, it appeared that the SC finally would be able to perform the role accorded to it in the ICRW. The United States had agreed to put its demands for zero quotas on hold; even Japan and the Soviet Union, still perhaps in shock from the hugely different environment they now faced in the commission, seemed more willing to cooperate in what promised to be a new era in cetacean management. Hopes of this nature, however, quickly evaporated as essentially the same uncertainty issues that had previously been raised in opposition to the SC's majority opinions again were put forward in order to undermine confidence in the NMP.

But the situation in the IWC by this point rather ironically already had been reversed. Uncertainty issues now were being raised in support of the moratorium by governments whose main priority was no longer the exploitation of the resource through sales of whale oil, but rather the exploitation of whales as endangered animals—and to some as sentient creatures—in order to realize the political benefits that protecting them offered. The SC, which had previously opposed the moratorium proposal, could no longer offer a consensus on either the NMP or the moratorium; the SC was divided between a majority who believed sustainable hunting, albeit at greatly reduced levels, was possible under the NMP and a minority who would only accept an end to all commercial hunts. This distinction is important because it illustrates how big a break this shift in perception represented: a veritable gestalt switch on the part of some scientists and governments concerning what the NMP's numerous uncertainty problems should be taken to mean.

In the past, the SC's divisions had been limited to questioning either the need for lower quotas or the extent to which quotas needed to be reduced; the minority opinion traditionally had argued for higher catches against a majority view that most often favored precautionary reductions. Agreement within the SC that some catches from some stocks were possible—except in cases such as with the blue and humpback stocks where the case for protection was very strong—had never before been in doubt, since none of the committee's scientists, regardless of their differences over catch limits, had considered a zero catch limit on all stocks of all species to be either justified or appropriate.

As the central argument of this study contends, the propensity among policy makers (and scientists) to demand greater levels of certainty is determined by the extent to which scientific advice describing possible outcomes conflicts with their priorities, as described by criteria I and II. Thus, the single most important determining factor behind the policies of the IWC has been the shifting political environment in which it has operated, rather than the uncertainty issues it has confronted. As a consequence, the ability of the SC to influence IWC policy has been severely limited by the management priorities of particular members or groups within the commission and their willingness to either invoke or ignore uncertainty issues in the pursuit of those priorities. This was the case with the five and then the two remaining Antarctic nations up to the late 1960s and early 1970s, and then later the like-minded majority within the commission that formed in support of the moratorium proposal during the late 1970s and early 1980s.

Some regime theory analysts, such as Haas, have suggested that advice from scientists and other "experts," based on "consensual knowledge," can influence policy making by providing policy makers with explanations of "cause and effect relationships" and the "complex interlinkages between issues" in response to uncertainty issues.[2] Within regime theory, the theoretical approach that has the most to say about the role of science in the construction and evolution of environmental regimes is that of cognitivists like Haas, who focus on the role of epistemic communities in fostering international cooperation on environmental problems.[3]

However, the example of the IWC clashes somewhat with cognitivist views on the importance of epistemic communities in providing scientific consensus as a basis for policy making. Cognitivists propose that because knowledge and ideas can alter the interests of governments, scientific convergence on the dimensions and effects of a given activity can lead to improved cooperation between states in managing its environmental impact. Examples such as international cooperation to reduce ozone depletion, which increased as uncertainty diminished (according to cognitivists), are often used to illustrate the influence scientific uncertainty can have on policy making. But as the IWC's experience indicates, scientific advice seldom comes without caveats and almost never is unanimously supported or unquestioned. How and why policy makers choose some advice over other advice in the absence of certainty or consensus, therefore, becomes an important question. Cognitivist theorists, by attributing a key role to scientific agreement in the successful functioning of a regime, appear to have largely ignored this question. Consequently, the ways in which such consensus can occur and the extent to which it needs to occur in order to have any meaningful impact both remain unclear.

I am not suggesting that the advice of scientists and other experts is of no significance in the formulation of policy. I am instead proposing a focus on the nature of science and how scientific advice is treated and judged by policy makers in order to better understand the role played by science in policy making, rather than simply contending that consensus among scientists or other groups of experts can (and does) influence and shape policies.

Haas acknowledges the epistemological problems associated with positioning science as a means of accessing and describing reality. At the same time, he bases the concept of epistemic communities and the role they play on the assumption that the collective efforts of epistemic communities can, over time, produce "consensus about the nature of the world."[4] One of the problems with this view is that such consensus often only occurs after previously disputed predictions of a particular result or threat are borne out by it actually taking place, as was the case with the Antarctic collapse. In this sense, the ability of epistemic communities to help avoid undesirable policy outcomes seems unclear, since advice of this nature is likely to be handicapped by the same kinds of political and economic priorities that have dominated IWC policy.

Because cognitivist theorists generally regard the resolution of scientific uncertainty issues to be an important determinant in regime policy formulation and cooperation, achieving scientific consensus becomes a critical factor in the success or failure of a regime. But in addition to the problem of how such consensus can occur, the closely related question of how much consensus is needed before uncertainty issues can be overcome is also left unanswered in any definitive sense.

As the IWC's experience with policy making—in addition to other environmental issues such as global warming—indicates, opinions among scientists or other expert groups are almost always divided. And even when a clear majority opinion has been available, as was the case in the IWC on several occasions, it seldom has been sufficient to compel those with important interests at stake to refrain from using lingering uncertainties as a basis for continued opposition. Examples of this include the Antarctic nations' refusals to adopt the recommendations of the COT and COF, the U.S. government's clear rejection of the SC's advice on the moratorium, and also the refusal of many in the commission to adopt the RMP on the unanimous advice of the SC in 1993.[5] As the IWC experience with scientific advice generally shows, consensus among scientists is a very rare thing. According to Butterworth,

> Post-1980 you'll find very few cases dealing with substantive issues upon which the Scientific Committee has given consensus advice. It is almost always that some advised this, but others advised that. Thus it is in a way

a bit misleading to say, for example, that the Scientific Committee never recommended the Antarctic Sanctuary ("contrary to the requirements of the Convention"). The Committee's modus operandi (arising originally from a desire to avoid voting) has led to a position where it will almost never say anything definitive on an issue of importance, because "lobby groups" from either side of the debate will be represented in the Committee and ensure the absence of a consensus statement contrary to their interests.[6]

Thus, while cognitivist theory recognizes the important role played by scientific uncertainty in regime policy, it fails to adequately address both the unavoidable existence and fundamental causes of the divisions over uncertainty issues that so often occur. Cognitivist theory is therefore unable to explain clearly what constitutes an epistemic community and why such communities of experts should be considered important in the removal of uncertainty as an impediment to cooperation on policy. M. J. Peterson, in applying an epistemic community–based analysis to the IWC, makes some questionable judgments when identifying epistemic community influence in the commission; he appears to set cohesiveness and uniformity of views among experts as a necessary precondition for epistemic communities.[7]

In Peterson's view, for example, it does not seem possible for one to be both a cetologist and an environmentalist.[8] This somewhat problematic assumption allows him to then cite the divisions within the IWC Scientific Committee as an important reason behind the committee's general inability to influence commission policy. Peterson writes, "The cetologists, the only group to qualify as an epistemic community [in relation to the IWC], suffered enough internal dissension and outright defections in the 1970s to lose that status temporarily." He goes on to argue, "It is likely that even a united cetologist community would have been unable to use the Scientific Committee as a mechanism for framing choices in the 1970s and early 1980s. The cetologists' internal divisions ensured that this became the case."[9]

But given the theory-dependent and unavoidably value-based nature of observing and interpreting "facts," how could it be possible to avoid divisions over the relevance or irrelevance of any uncertainties associated with the data at hand? And as Peterson himself notes, even unanimously supported scientific advice was unlikely to have given the SC any more influence, as has been the case on the very few occasions when the committee did unanimously support its recommendations. The very nature of science, and the problems it involves, means that uncertainty is unavoidable in any research.

Scientific advice, therefore, is always open to attack by either (a) invocation of uncertainty to argue the evidence is insufficient to conclude harm will not occur (i.e., the precautionary principle in its current form), or (b) invocation of uncertainty based on a lack of scientific evidence to say harm will occur (i.e., the inverse of the precautionary principle in its current form). And as the IWC's experience with uncertainty issues and problems suggests, these interpretations are applied based on the political/economic interests of the parties involved and the degree to which these often divergent interests prevent the establishment of commonly accepted goals and priorities.

A major obstacle to cooperation on whale management, I propose, has been the perception among policy makers and scientists in general, and the public at large, of what science represents and what scientific advice can realistically be expected to provide. If science is commonly expected to accurately describe and predict the real world (as per the conventional view of science), then it is not difficult for governments and organizations to selectively criticize and reject scientific advice when it conflicts with their needs on the grounds of uncertainty—the case with the Dutch delegation during the fin whale debate, Soviet and Japanese resistance to quota reductions in the mid to late 1960s, and the U.S. position on the NMP. But if it is accepted from the outset, for the reasons discussed in the introduction, that uncertainty must be dealt with by making political decisions that concern acceptable risk (because certainty is beyond the grasp of science), it becomes more difficult for participants to hide political and economic interests behind ostensibly scientific arguments.

On the few occasions when relatively high levels of certainty over the depletion of stocks existed, such as the Antarctic collapse, the Japanese and Soviet delegations were quick to point to the increasing hardships being suffered by their industries in order to bolster their positions against reductions, but only after it had become clear that uncertainty arguments alone no longer would suffice. This pattern of shifting justifications when uncertainty arguments become weaker in the face of contradictory evidence or circumstances was again repeated in the early 1990s, this time by anti-whaling governments and groups trying to reorganize the basis of their opposition in response to the decreased risk offered by the RMP. In both instances, the erosion of uncertainty as a basis for opposition—resulting from the willingness of the majority to *believe* the SC's interpretation of the data—helped to reveal the underlying political reasons for the policies being pursued.

Indeed, the strategy of arguing against a preventive course of action because we do not know or cannot prove that a particular set of outcomes will occur is common enough, as is the more recent strategy of arguing against other courses of action because we cannot prove undesirable outcomes will *not* occur. But both of these strategies become difficult to implement when certainty is recognized from the outset to be unavailable—a situation that reveals the political interests at the heart of the issue at hand. Put another way, uncertainty only need be a problem when people believe they can achieve their goals by demanding more certainty. The actions during the 1950s of all the IWC's members—except for New Zealand, which agreed with the SC's advice to reduce the Antarctic quota to 11,000 BWU—clearly reflected this style of reasoning.

While the Dutch delegation was by far the most prolific employer of scientific uncertainty in order to justify its policies in the 1950s and early 1960s, the Dutch certainly were not alone in this regard. The British commissioner relied on scientific uncertainty as an argument against further quota reductions at the Moscow meeting, indicating that he still was not convinced of the dangers facing the Antarctic stocks. Other governments also balked at conservation initiatives that may have handicapped the various whaling companies' needs to cover spiraling costs. Examples of this include keeping open the Antarctic Sanctuary beyond the three years recommended by the Scientific Sub-Committee (which had the general support of the commission),[10] the widespread opposition to the SC proposal to protect North Atlantic and North Pacific blue whales after 1955,[11] and also the general unwillingness of the Antarctic nations to compromise over their individual shares of the proposed Antarctic quota—thereby causing negotiations to fail.

In effect, the Antarctic whaling members, who largely dominated the IWC's policies during this period, were applying the central tenet of what would become recognized as the precautionary principle in reverse. Rather than stipulating that uncertainty should not be used to prevent measures being enforced to avoid the collapse of the Antarctic stocks, they were instead arguing that advice involving uncertainty should not be accepted as a basis for policies that could damage the industry. Ironically, both positions are essentially precautionary in their intent with the only major difference being the designated focus of the intended protection (i.e., the resource or the short-term interests of the industry). Over the next two decades, however, the significant change in the IWC's focus and priorities that slowly occurred, and the effects this change had upon perceptions of scientific advice and uncertainty within the commission, set the stage for the moratorium's adoption in 1982.

The ongoing tendency within the IWC to distinguish between science and politics, and to assume that science exists in some kind of vacuum without political influence, has only further obscured the various scientific uncertainty debates. This attitude has denied the unavoidable link between science and politics, thereby encouraging focus on disputes over who has the "real science" rather than on the more relevant issue of which science is politically acceptable. Therefore, it is not simply scientific uncertainty but rather the commission's failure to openly address the political reasons underpinning the application of scientific uncertainty that has stalled the RMP's implementation.

The emergence of the precautionary principle as the guiding tenet for IWC policy was positive in that it drew attention to the need for scientific uncertainty to be allowed for in the design of cetacean management regimes and it led to more research. Indeed the IWC's application of the precautionary principle—in the form of the comprehensive assessment, the RMP, and the observation and inspection provisions included under the RMS—has been based on a common belief that scientific uncertainty should be accounted for in the management of cetaceans. But the IWC has not made any serious attempts to clarify *how* to manage scientific uncertainty—in terms of facilitating political agreement among members over how much (if any) scientific uncertainty is acceptable—or how to answer the crucial question of to what ends scientific uncertainty should be managed. As one long-serving member of the IWC Scientific Committee has observed, the moratorium continues to exist more in spite of science rather than because of it:

> Do I think in retrospect that the adoption of the moratorium was justified, on the grounds of prevailing uncertainties? Well, in the sense that the adoption of the moratorium forced the development of the RMP, which directly addressed scientific uncertainties in assessment, then I think it was justified. On the other hand, in the sense that the moratorium has now become what we all feared it might, an indiscriminate and permanent ban on whaling, then I don't think it was justified.[12]

The generally recognized precautionary principle caveat that a lack of full scientific certainty should not delay measures to prevent environmental damage appears to have been interpreted by many in the IWC to mean that commercial whaling can only resume with full certainty of no serious danger to stocks—even though the commission instigated work on the comprehensive assessment assuming that no such certainty is available. And by the time the IWC completed and adopted the RMP, some members had even gone so far as to say that scientific uncertainty was not even relevant to the question of whether or not to recommence commercial whaling. In its opening statement to the IWC's 1994 meeting, New Zealand stated, "We will work to maintain the moratorium on commercial whaling because it reflects the current reality of world opinion. We will participate fully in the dialogue about the Revised Management Scheme because it is essential to have the best possible rules for whaling, whether or not they are required in practice."[13]

The important role played by scientific uncertainty in the IWC, I believe, supports my contention that it is largely the competition between the perceived desires or needs of the various actors, rather than simply theories and claims to knowledge, that actually determines policy choices.[14] The main issue is not whether we can say who is using the real science and who is attempting to distort it, but rather how science is and can be used in policy making and, perhaps most importantly, to what ends it is used. For the reasons I have outlined in the introduction, I do not believe it is possible to explain why we accept some theories and not others purely in terms of which better explains "reality," not least because reality is itself a controversial subject. And for this reason, I do not think it is helpful to explain the IWC's problems only in terms of who is or is not basing their position on "the best scientific advice" or only by treating consensus on management advice as something that can be expected to occur at some point.

However, I must emphasize that I am not arguing against citing empirical observations and methods as a reason for doing or not doing something (or for supporting particular statements about the nature of the world). What I am arguing is that we must understand the limitations involved and the important caveat that it is most often our desires and needs that, at the end of the day, will have most influenced our choices. As the IWC experience clearly demonstrates, science is not value-neutral and cannot, therefore, operate independently of the larger political environment in which it exists.

Consideration of this inherently political aspect of science, and the important role it plays in explaining the treatment of science in policy making, is an essential part of understanding why scientists interpret data differently and often disagree over their conclusions. Such disagreements cannot be accounted for only in terms of whose observation and interpretation of the facts is right or wrong, since it is not possible to prove any set of interpretations and conclusions to be true. It is rather the differing values and assumptions that necessarily guide any scientist in his or her work that can explain not only why such stand-offs occur, but that also can shed light on of how policy makers choose some advice and reject other advice without the benefit of *knowing* which may or may not be correct.

Thus, while I would argue, as I have done here, that scientific uncertainty issues have been an important factor in shaping the IWC's various policy choices, I also would argue that science and its various shortcomings have never been the real issue behind the IWC's various disputes and policy decisions. Instead, I am proposing that political agendas are what set the parameters of policy making (and to a large degree of science itself); therefore, scientific uncertainty and the precautionary principle, in practice, often have been treated by many policy makers as little more than tools to in the pursuit of their political objectives. In other words, it is not uncertainty itself that determines or influences policy making so much as how we choose to use it—and that is ultimately determined by political choices about what is or is not desirable.

APPENDIX

TABLE 1. Antarctic pelagic whaling (1945–78) quotas, seasons, catches, and fleets (not including revisions for falsified Soviet catch data)

Pelagic Antarctic season	Total quota of BWU set by IWC	Total catch of BWU	Average catch of BWU catch per catcher day	Floating factories	Catchers
1945–46	16,000	7,310.0	0.89	9	77
1946–47	16,000	15,304.2	1.06	15	129
1947–48	16,000	16,364.3	0.92	17	162
1948–49	16,000	16,007.4	0.84	18	191
1949–50	16,000	16,062.1	0.88	18	216
1950–51	16,000	16,416.2	0.86	19	239
1951–52	16,000	16,007.7	0.94	19	263
1952–53	16,000	14,866.6	0.86	16	230
1953–54	15,500	15,456.4	0.98	17	206
1954–55	15,500	15,323.5	0.91	19	233
1955–56	15,000	14,874.3	0.99	19	257
1956–57	14,500	14,745.2	0.95	20	225
1957–58	14,500	14,850.9	0.90	20	237
1958–59	15,000	15,300.8	0.94	20	235
1959–60	17,500 [a]	15,511.7	0.73	20	220
1960–61	17,780 [a]	16,433.5	0.68	21	252
1961–62	17,780 [a]	15,252.6	0.51	21	261
1962–63	15,000	11,306.1	0.50	17	201
1963–64	10,000	8,429.0	0.41	15	180
1964–65	8,000 [a]	6,986.1	0.40	15	172
1965–66	4,500	4,090.9	0.31	10	128
1966–67	3,500	3,511.8	0.30	9	120

TABLE 1. *(continued)*

Pelagic Antarctic season	Total quota of BWU set by IWC	Total catch of BWU	Average catch of BWU catch per catcher day	Floating factories	Catchers
1967–68	3,200	2,803.7	0.29	8	97
1968–69	3,200	2,472.6	0.30	6	84
1969–70	2,700	2,477.2	0.30	6	84
1970–71	2,700	2,470.5	0.30	6	86
1971–72	2,300	2,250.8 [b]		6	84
1972–73	1,808	1,524.5 [c]		6	78
1973–74	1,475	1,376.0 [d]		6	75
1974–75	1,167 [e]	1,132.7 [e]		6	72
1975–76	482 [f]	406.5 [f]		5	56
1976–77	310 [g]	310.0 [g]		4	43
1977–78	128.5 [h]	95.0 [h]		3	34
1978–79	—[i]	—[i]		3	31

SOURCE: J. N. Tonnessen and A. O. Johnsen, *The History of Modern Whaling* (Berleley: University of California Press, 1982), 749–50. Adapted with permission of University of California Press.

NOTES: [a] Seasons where no Antarctic quota was set by the IWC. Figures shown represent the totals of the individual limits agreed to by each Antarctic whaling nation for the season indicated; [b] Including 3,021 minke whales; [c] Including 5,745 minke whales; [d] Including 7,713 minke whales; [e] Including 7,000 minke whales; [f] Including 6,034 minke whales; [g] Sei whales only (1,858) converted to BWU; [h] Sei whales only (771) converted to BWU, plus 5,690 minke whales; [i] Only minke whales (5,446) were allowed to be caught.

TABLE 2. Average prices of baleen whale oil (1900–77)

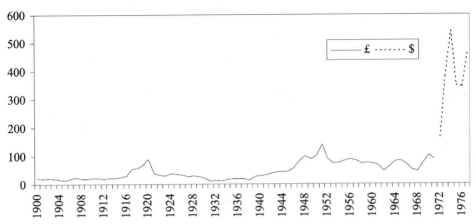

SOURCE: J. N. Tønnessen and A. O. Johnsen, *The History of Modern Whaling* (Berkeley: University of California Press, 1982), p. 753. Adapted with permission of University of California Press.

NOTES: The figures are the middle-prices of the highest and lowest annual sale. Price per long ton = 1,016 kg = 6 barrels. From 1920 (et seq.): the price is that of the Antarctic production of the previous season.

192 APPENDIX

TABLE 3. Catches of whales in the Antarctic (1919–20 to 1976–77)
(not including revisions for falsified Soviet catch data)

Year	Blue	Fin	Humpback	Sei-Bryde's	Minke	Sperm	Others	Total
1919–20	1,874	3,213	261	71	—	8	14	5,441
1920–21	2,617	5,491	260	36	—	31	13	8,448
1921–22	4,416	2,492	9	103	—	3	—	7,023
1922–23	5,683	3,677	517	10	—	23	—	9,910
1923–24	3,732	3,035	233	193	—	66	12	7,271
1924–25	5,103	4,366	359	1	—	59	—	10,488
1925–26	4,697	8,916	364	195	—	37	10	14,219
1926–27	6,545	5,102	189	778	—	39	12	12,665
1927–28	8,334	4,459	23	883	—	72	4	13,775
1928–29	12,734	6,689	48	808	—	62	—	20,341
1929–30	17,898	11,614	853	216	—	73	1	30,665
1930–31	29,410	10,017	576	145	—	51	2	40,201
1931–32	6,488	2,871	184	16	—	13	—	9,572
1932–33	18,891	5,168	159	2	—	107	—	24,327
1933–34	17,349	7,200	872	—	—	666	—	26,087
1934–35	16,500	12,500	1,965	266	—	577	—	31,808
1935–36	17,731	9,697	3,162	2	—	399	—	30,991
1936–37	14,304	14,381	4,477	490	—	926	2	34,579
1937–38	14,923	28,009	2,079	161	—	867	—	46,039
1938–39	14,081	20,784	883	22	—	2,585	2	38,356
1939–40	11,480	18,694	2	81	—	1,938	8.6	32,900
1940–41	4,943	7,831	2,675	110	—	804	—	16,363
1941–42	59	1,189	16	52	—	109	—	1,425
1942–43	125	776	—	73	—	24	—	998
1943–44	339	1,158	4	197	—	101	—	1,799
1944–45	1,042	1,666	60	78	—	45	—	2,891
1945–46	3,606	9,185	238	85	—	273	—	13,387
1946–47	9,192	14,547	29	393	—	1,431	1	25,593
1947–48	6,908	21,141	26	621	—	2,622	—	31,318
1948–49	7,625	19,123	31	578	—	4,078	—	31,435
1949–50	6,168	18,061	2,117	101	—	2,570	—	29017
1950–51	6,966	17,474	1,630	367	—	4,742	—	31,179
1951–52	5,124	20,520	1,546	32	—	5,344	—	32,566
1952–53	3,866	21,197	954	123	—	2,185	—	28,325
1953–54	2,684	24,986	594	251	—	2,700	—	31,215
1954–55	2,163	25,878	493	146	—	5,708	—	34,388
1955–56	1,611	25,289	1,432	273	—	6,881	—	35,489

TABLE 3. *(continued)*

Year	Blue	Fin	Humpback	Sei-Bryde's	Minke	Sperm	Others	Total
1956–57	1,505	25,700	679	712	—	4,345	—	32,941
1957–58	1,684	25,222	396	2,385	—	6,310	—	35,997
1958–59	1,191	25,837	2,394	1,402	—	5,437	—	36,261
1959–60	1,230	26,415	1,338	3,234	—	4,138	—	36,355
1960–61	1,740	27,374	718	4,310	—	4,666	—	38,810
1961–62	1,118	27,099	309	5,196	—	4,829	2	38,552
1962–63	947	18,668	270	5,503	—	4,771	1	30,159
1963–64	112	14,422	2	8,695	—	6,711	—	29,942
1964–65	20	7,811	—	20,380	—	4,352	—	32,563
1965–66	1	2,536	1	17,587	—	4,555	—	24,680
1966–67	4	2,896	—	12,638	—	4,960	—	20,495
1967–68	—	2,155	—	10,357	—	2,568	—	15,080
1968–69	—	3,020	—	5,776	—	2,682	—	11,478
1969–70	—	3,002	—	5,867	—	3,090	—	11,959
1970–71	—	2,890	—	6,153	—	3,055	—	12,098
1971–72	—	2,683	3	5,456	3,021	3,366	—	14,529
1972–73	7	1,761	5	3,864	5,745	4,203	—	15,585
1973–74	—	1,288	—	4,392	7,713	4,927	—	18,320
1974–75	—	979	—	3,859	7,000	4,162	—	16,000
1975–76	—	206	—	1,821	6,034	2,829	—	10,890
1976–77	—	—	—	1,858	7,900	2,002	—	11,760
Total	307,370	640,357	35,435	139,407	37,143	136,177	779	1,296,938

SOURCE: S. A. Mizroch, "The Development of Balaenopterid Whaling in the Antarctic," *Cetus*, vol. 5, no. 2 (1984), quoted in W. Aron, "Science and the IWC," in R. L. Friedheim, ed., *Toward a Sustainable Whaling Regime*, 112–13 (Seattle: University of Washington Press, 2001). Adapted here with permission of University of Washington Press.

TABLE 4. Catches of Whales in the Antarctic (1948–49 to 1971–72)
(including revisions for falsified Soviet catch data)

Year	Blue	Fin	Humpback	Sei-Bryde's	Minke	Sperm	Other	Total
1948	7,624	19114	143	579	—	4,000	13	31,473
1949	6,182	19763	2,618	1,284	—	2,706	2	32,555
1950	6,976	19219	2,583	633	—	4,933	1	34,345
1951	5,042	22739	1,976	535	4	5,388	19	35,703
1952	3,833	22779	1,157	626	6	2,606	10	31,017
1953	2,697	27684	708	964	3	2,820	12	34,888
1954	2,170	28379	1,698	525	—	5,666	38	38,476
1955	1,585	28064	4,145	495	33	6,891	38	41,251
1956	1,501	27555	954	1,610	45	4,492	68	36,225
1957	1,660	26847	2,571	3,138	11	6,265	32	40,524
1958	1,076	26460	6,013	2,365	11	5,206	3	41,134
1959	841	24542	13,563	3,471	2	3,481	78	45,978
1960	1,603	26232	12,945	4,965	3	4,338	6	50,092
1961	992	26224	5,546	5,611	3	6,522	1,365	46,263
1962	1,724	17286	2,932	5,644	21	7,040	729	35,376
1963	1,469	13778	693	9,486	116	13,392	383	39,317
1964	3,228	7514	234	21,259	15	13,623	86	45,959
1965	1,019	3123	2,060	19,712	360	7,018	20	33,312
1966	424	3222	1,035	16,683	32	7,762	206	29,364
1967	396	2546	929	15,949	621	6,008	12	26,461
1968	674	3492	2	10,480	54	5,441	—	20,143
1969	920	2896	—	9,915	37	7,808	88	21,664
1970	833	8353	1,700	8,524	42	1,334	12	20,798
1971	537	2233	3	7,111	3,060	12,211	7	25,162
1972	7	1761	5	3,864	5,745	8,743	42	20,167
Total	55,013	411,805	66,213	155,428	10,224	155,694	3,270	857,647

SOURCE: IWC records. Adapted with permission of IWC.
NOTE: "Other" includes right whales, other large whales not mentioned, and small whales.

TABLE 5. Comparison of old and new catch data for Sovietskaya Ukraina

Dates	Blue	Pygmy blue	Sperm	Fin	Humpback	Sei	Bryde's	Minke	Right	Other	Total
New	68	0	230	31	7,520	236	0	0	78	0	8,163
Old	428	0	1,735	604	440	915	0	184	0	44	4,350
(7 November 1959–10 April 1960)											
New	47	0	625	239	5573	567	0	2	0	0	7,053
Old	158	0	1,607	608	120	753	0	135	0	33	3,414
(1 November 1960–6 April 1961)											
New	75	0	1,584	188	1,078	761	0	0	1,314	1	5,001
Old	253	0	1,660	547	125	702	0	0	1	0	3,288[a]
(7 November 1961–6 April 1962)											
New	105	0	1,236	883	667	642	4	16	254	0	3,807
Old	110	0	1,267	980	188	602	0	16	0	0	3,163
(1 November 1962–16 April 1963)											
New	12	40	1,072	1,534	299	1,680	0	0	283	15	4,935
Old	0	0	860	1,435	0	1,440	0	5	0	10	3,750
(5 November 1963–22 April 1964)											
New	0	1,818	239	2,568	3	1,090	465	2	59	0	6,244
Old	0	0	605	2,143	0	1,785	0	0	0	0	4,533
(30 October 1964–24 May 1965)											
New	106	312	673	1,401	710	1,357	92	7	48	10	4,716
Old	0	0	377	1,373	0	909	0	7	0	7	2,673
(14 November 1965–25 April 1966)											

TABLE 5. (continued)

Dates	Blue	Pygmy blue	Sperm	Fin	Humpback	Sei	Bryde's	Minke	Right	Other	Total
New	66	46	1,048	1,735	483	1,685	35	8	17	4	5,127
Old	0	0	644	1,689	0	383	3	5	0	3	2,727[b]
(5 November 1966–7 May 1967)											
New	9	310	?	1,867	?	2,217	?	1	?	?	5,059
Old	0	0	351	1,116	0	1,115	0	1	0	3	2586[c]
(10 November 1967–28 April 1968)											
New	65	21	?	?	?	3,051	?	?	?	?	5,283
Old	0	0	480	1,153	0	713	0	11	0	0	2360[d]
(11 November 1968–25 April 1969)											
New	544	?	?	1,734	?	?	?	?	?	?	6,721
Old	0	0	391	1,650	0	1,000	0	0	0	0	3,041[e]
(7 November 1969–29 April 1970)											
New	0	0	?	?	?	?	?	?	?	0	5,438
Old	0	0	553	1,654	0	640	0	5	1	1	2859[f]
(1November 1970–20 April 1971)											
New	5	71	337	4,088	?	?	?	?	?	?	6,228
Old	0	0	547	1,966	0	452	0	14	0	0	2979[g]
(1 November 1971–6 May 1972)											

TABLE 5. (continued)

Total catch summary (new and old) 1959–60 to 1971–72

Dates	Blue	Pygmy blue	Sperm	Fin	Humpback	Sei	Bryde's	Minke	Right	Other	Total
New (1959–60 to 1971–72)	558	3,162	7,044	16,268	16,333	13,286	596	36	2,053	30	73,775
Old	949	0	11,077	16,921	873	11,409	3	383	1	105	41,723

SOURCE: IWC, *Forty-fifth Report*, 1995, 133; and IWC, *Forty-sixth Report*, 1996, 131–38. Reproduced here with permission of IWC.

NOTES: No new data were available for the period 1972–73 to 1986–87, but since national inspectors were on board at this time, the data are believed to be more reliable. Adjustments have been made to these new Soviet totals as there were discrepancies in the sums. There may be additional catches to add to the numbers below (indicated by a "?" in the annual catch data). [a] Incl. Sci. Permit: 1 Right; [b] Incl. lost: 1F, 2Sei, Sci. Permit 3Bryde's; [c] Incl. lost: 1F, 2Sei; [d] Incl. lost: 3F, 2Sei; [e] Incl. lost: 1F, 2Sp, 3Sei; [f] Incl. lost: 1Sei, 2F, 1SP, Sci. Permit: 2Pri; [g] Incl. lost: 4SP.

TABLE 6. Participation in IWC meetings (1965–95)

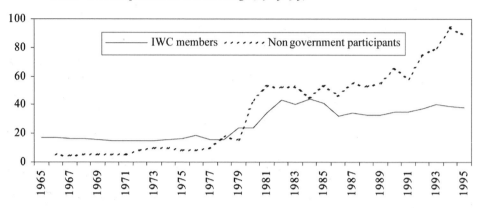

SOURCE: R. L. Friedheim, "Negotiating in the IWC Environment," in *Toward a Sustainable Whaling Regime* 212 (Seattle: University of Washington Press, 2001); and IWC annual reports. Reproduced with permission of University of Washington Press.

APPENDIX 199

MAP A. Map of world showing Antarctic areas and (dotted) regions closed to factory ships for the purpose of taking and treating whales

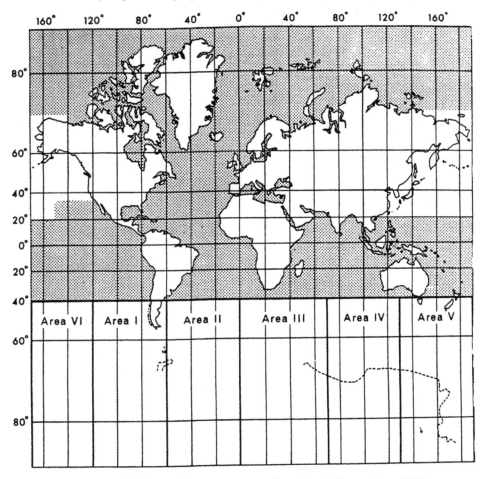

SOURCE: IWC *Sixteenth Report,* 1966, p. 73. Reproduced with permission of IWC.

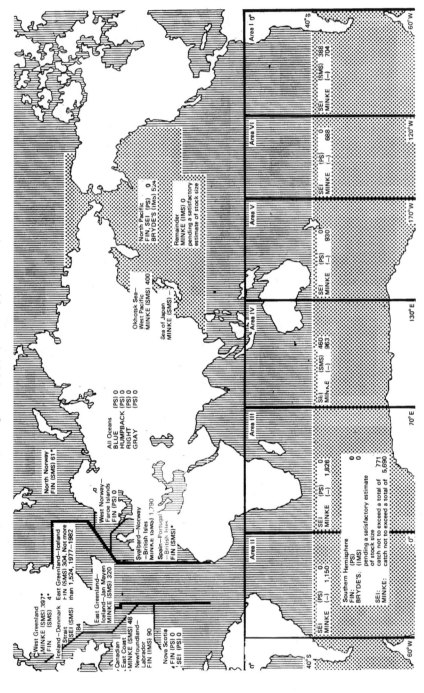

MAP B. Baleen whale stocks, classification, and quotas (1977–78 and 1978 seasons)

SOURCE: IWC, *Twenty-eighth Report*, 1978. Reproduced with permission of the IWC.

NOTES: * provisional; (PS) Protection Stock; (SMS) Sustained Management Stock; (IMS) Initial Management Stock; (—) not classified.

MAP C. Baleen whale (excluding Bryde's whale) stocks, classification, and catch limits (1982–83 and 1983 seasons)

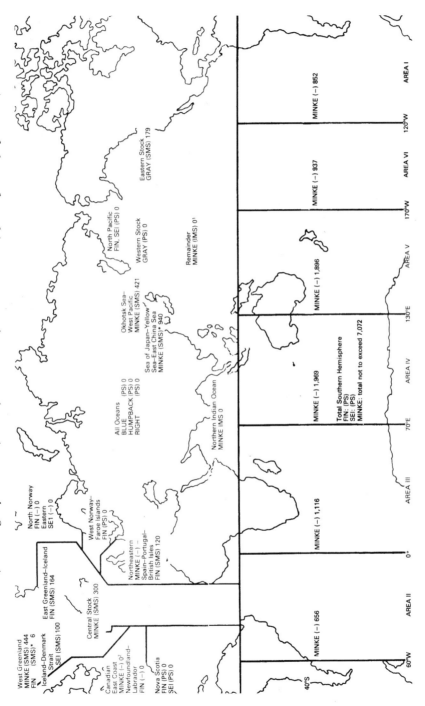

SOURCE: IWC, *Twenty-third Report*, 1978. Reproduced with permission of the IWC.

NOTES: * provisional; (PS) Protection Stock; (SMS) Sustained Management Stock; (IMS) Initial Management Stock; (–) not classified.

NOTES

1 / INTRODUCTION

1. The term "commercial whaling" was not used in the IWC until 1972, when copies of a communication from that year's UN Conference on the Human Environment calling for "a ten year moratorium on commercial whaling" were distributed at an IWC meeting. At the same meeting, the United States proposed that such a moratorium be adopted, which the United Kingdom seconded. The introduction of the distinction between commercial and aboriginal whaling was an important one, and it fueled considerable debate in the IWC over the following decades. The U.S. government has used the distinction on many occasions to oppose whaling operations outside the United States, while simultaneously supporting the rights of indigenous whalers in the country to continue limited hunting of often endangered species. The United States also used this distinction to fend off international criticism over the large number of small cetaceans (porpoises and dolphins) killed by the U.S. tuna fishery during the 1970s, claiming that these deaths were not commercial but only "incidental." Another perspective on the distinction has been provided by Junichi Takahashi, who argues the commercial/aboriginal whaling distinction was introduced to gain political advantage for the primarily Western-based opponents of whaling, many of whom are English speaking, by utilizing culturally biased English definitions of "commercial" and "aboriginal." See J. N. Tønnessen and A. O. Johnsen, *The History of Modern Whaling* (Berkeley: University of California Press, 1982), 674–77 (an abridged English translation of the original, Norwegian four-volume work entitled *Den Moderne Hvalfangsts Historie*; the original work contains many tables and footnotes that are not included in the English edition); and J. Takahashi, "English Dominance in Whaling Debates: A Criti-

cal Analysis of Discourse in the International Whaling Commission," *Nichibunken Japan Review*, vol. 10 (1998).

2. See Tønnessen and Johnsen, *History of Modern Whaling*,. 667–72; J. E. Scarff, "The International Management of Whales, Dolphins, and Porpoises: An Interdisciplinary Assessment (Part One)," Ecology Law Quarterly, vol. 6 (1977): 366–71.

3. Whales, generally classified as Cetacea, can be separated into two main groups: baleen (Mysticeti) and toothed whales (Odontoceti). There are ten species of baleen and sixty-five species of toothed whales (the largest being the sperm whale). The three families of baleen whale are: (a) the medium-sized and slow-swimming right whales; (b) the gray and humpback whales; and (c) the mostly larger and faster-swimming rorqual whales, which includes the blue, fin, sei, Bryde's whale, and the much smaller minke—all of which belong to the genus *Balaenoptera*. The larger right whales inhabit only the northern hemisphere and are known as the Greenland right whale or the closely related bowhead whale, while the smaller species, the southern right whale, inhabits only the southern hemisphere. Gray whales inhabit only the northern hemisphere, but humpbacks are found worldwide—both of these whales also are medium-sized and are relatively slow swimming. See K. R. Allen, *Conservation and Management of Whales* (Seattle: University of Washington Press; London: Butterworths, 1980), 2–3.

4. Sperm-whale oil is actually more like a form of liquid wax and is often referred to as "sperm oil" to distinguish it from the edible whale oil taken from the baleen whales. Thus, sperm oil was used only for lighting and as a high-grade machine oil. The meat of the sperm whale, unlike that of the baleen whales, also was not favored for consumption by the Japanese, although some meat was sold on the domestic market. Kay Radway Allen has suggested that the adoption of mineral oil over sperm oil for lighting and so on that occurred after the discovery of oil in Texas in the mid-1800s may account for the decline in sperm-whale catches at this time. Allen adds, however, that the situation is unclear and also could have been caused by depletion of stocks. See Allen, *Conservation and Management of Whales*, 12 and 16; Tønnessen and Johnsen, *History of Modern Whaling*, 7 and 228; G. L. Small, *The Blue Whale* (New York: Columbia University Press, 1971), 96.

5. See for example, Small, *Blue Whale*; P. Birnie, *International Regulation of Whaling: From Conservation of Whaling to Conservation of Whales and Regulation of Whale Watching*, vols. 1 and 2 (New York: Oceana Publications, 1985); M. M. R. Freeman, "Why Whale?" in O. D. Jonsson, ed., *Whales and Ethics*, 39–56 (Reykjavik: Fisheries Research Institute, University of Iceland University Press, 1992); P. Stoett, *The International Politics of Whaling* (Vancouver: Uni-

versity of British Columbia Press, 1997); T. Schweder, *Intransigence, Incompetence or Political Expediency? Dutch Scientists in the International Whaling Commission in the 1950s: Injection of Uncertainty*, IWC Scientific Committee Paper SC/44/o 13, 1992; M. J. Peterson, "Whalers, Cetologists, Environmentalists, and the International Management of Whaling," in P. M. Haas, ed., *Knowledge, Power, and International Policy Coordination*, 147–86 (Columbia: University of South Carolina Press, 1997); G. Trigg, "Japanese Scientific Whaling: An Abuse of Right or Optimum Utilisation?" *Asia Pacific Journal of Environmental Law*, vol. 5, no. 1 (2000): 33–59; D. Goodman, "If Only Greenpeace Told the Truth about Whaling," *The Japan Times*, March 16, 2000; L. Gregoriadis, "Norwegian Fleet Sets Off to Kill 750 Minke in Biggest Slaughter of Whales for 14 Years," *The Independent View from Europe* (*Daily Yomiuri* supplement), May 9, 1999; and C. Lord, "Deep Feeling," *The Courier-Mail*, April 25, 2000.

6. For a discussion of the dominance of realist approaches to international relations theory and the emergence of regime theory as a means of reconciling the differences between realism and interdependence-based theories, see S. Haggard and B. A. Simmons, "Theories of International Regimes," *International Organisation*, vol. 41, no. 3 (1987): 491–517. For realist perspectives on interdependence and the role of regimes in international relations theory see R. O. Keohane and J. S. Nye, "Power and Interdependence Revisited," *International Organization*, vol. 41, no. 4 (1987): 725–53; and J. M. Grieco, "Anarchy and the Limits of Co-operation: A Realist Critique of the Newest Liberal Institutionalism," *International Organization*, vol. 42, no. 3 (1988): 485–508. For discussions of the factors required for regime formation, see O. R. Young and G. Osherenko, eds., *Polar Politics: Creating International Environmental Regimes* (Ithaca: Cornell University Press, 1993).

7. See P. M. Haas, *Knowledge, Power, and International Policy Co-ordination*; and "Do Regimes Matter? Epistemic Communities and Mediterranean Pollution Control," *International Organization*, vol. 43, no. 3 (1989): 377–403.

8. M. Crenson, "Particle Collider to Re-create Birth of Universe," *Daily Yomiuri*, March 28, 2000, 10.

9. A. F. Chalmers, *What Is This Thing Called Science?* (Saint Lucia: University of Queensland Press, 1999), xix (my italics).

10. International Convention for the Regulation of Whaling, Article V, in IWC, *First Report*, 1950, appendix 1.

11. See K. R. Popper, *The Logic of Scientific Discovery* (London: Hutchinson, 1972); and T. S. Kuhn, *The Structure of Scientific Revolutions* (Chicago: University of Chicago Press, 1996). For an excellent and readable discussion of demarcation between scientific and nonscientific inquiry, its associated problems,

and treatment by Popper, Kuhn, Lakatos, and Feyerabend, see B. Larvor, *Lakatos: An Introduction* (New York: Routledge, 1998); see also Chalmers, *What Is This Thing Called Science?* For the main points of Feyerabend's arguments against scientific method and demarcation, see J. Preston, ed., *Paul K. Feyerabend: Knowledge, Science and Relativism* (Cambridge: Cambridge University Press, 1999); and P. K. Feyerabend, "On the Critique of Scientific Reasoning," in R. S. Cohen, P. K. Feyerabend, and M. W. Wartofsky, eds., *Essays in Memory of Imre Lakatos*, 109–43 (Dordrecht, Germany: D. Reidel Publishing Company, 1976).

12. Larvor, *Lakatos*, 91.

13. This is a common argument made against scientific method and epistemology in general by postmodernist critics. See for example Barry Hindess's unrestrained attack on Popper in chapter 6 of B. Hindess, *Philosophy and Methodology in the Social Sciences* (Sussex: The Harvester Press, 1977).

14. E. H. Carr, *What is History?* (Oxford: Oxford University Press, 1961), 131.

15. Chalmers, *What Is This Thing Called Science?* 41–58.

16. B. Russell, *Sceptical Essays* (New York: Barnes and Noble, 1961), 28–29 (including quote from Whitehead's *Science and the Modern World* [1926]).

17. Quoted in Chalmers, *What Is This Thing Called Science?* 66–67.

18. Hindess, *Philosophy and Methodology in the Social Sciences*, 179.

19. Kuhn, *Structure of Scientific Revolutions*, 146–47.

20. Ibid., 111–14.

21. Larvor, *Lakatos*, 62–72.

22. Ibid., 50–57.

23. Ibid.

24. Kuhn, *Structure of Scientific Revolutions*, 205–7.

25. Z. Sardar, *Thomas Kuhn and the Science Wars* (Cambridge: Icon Books, 2000), 63.

26. Russell, *Sceptical Essays*, 30.

27. Ibid., 30–31.

28. Bentham, quoted in J. H. Burns and H. L. A. Hart, eds., *An Introduction to the Principles of Morals and Legislation* (London: The Athlone Press, 1970), 11–12.

29. Ibid., 13.

30. See G. Dworkin, "Paternalism," in G. Dworkin, ed., *Mill's On Liberty: Critical Essays*, 61–82 (Lanham, MD: Rowman and Littlefield Publishers, 1997).

31. L. Doyal and I. Gough, *A Theory of Human Need* (London: Macmillan, 1991), 43.

32. Ibid., 45.

33. Ibid., 73.

34. Ibid., 97.

35. Many of the questions and issues raised by the philosophy of science debate over what science and the research it generates represent are also present in an almost identical debate among historians, most of whom fit into either the relativist or positivist camp in terms of what they believe historians produce when they attempt to write a history of some event or period in order to solve a given question or problem. One of the first shots in the ongoing war of ideas between the relativists and positivists was fired by Charles Beard in his 1933 criticism of fellow American Historical Association member Theodore Clarke Smith's claim that historians pursued "the ideal of the effort for objective truth." According to James T. Kloppenberg, Beard replied by saying that "'the noble dream' of scientific objectivity was but an illusion. Writing history, Beard insisted, is 'an act of faith.' Despite all efforts to know the past, the historian 'remains human, a creature of time, place, circumstances, interests, predilections, culture,' characteristics that shape historical work and are inevitably reflected in it." J. T. Kloppenberg, "Objectivity and Historicism: A Century of American Historical Writing," *American Historical Review*, vol. 94, no. 4 (1989): 1011–12. Kloppenberg provides a good overview of the positivist/relativist debate among historians.

36. See B. Greenwood, "Between Hope and a Hard Place," *Nature*, vol. 430 (August 2004): 926–27.

37. Address by T. Stamps at the closing session of the Third Pan-African Malaria Conference, Nairobi, Kenya, June 1998. See also E. Ruppel Shell, "Resurgence of a Deadly Disease," *Atlantic Monthly*, August 1997, 45–60.

38. Australian Department of the Environment and Heritage, "A Universal Metaphor: Australia's Opposition to Commercial Whaling," report of the National Task Force on Whaling, May 1997, http://www.deh.gov.au/coasts/species/cetaceans/whaling/.

39. See, for example, S. J. Heishman, "Behavioral-Cognitive Effects of Smoking," paper presented at Addicted to Nicotine: A National Research Forum, Bethesda, MD, July 27–28, 1998, http://www.drugabuse.gov/meetsum/nicotine/heishman.html.

2 / THE IWC 1949–1959

1. K. R. Allen, *Conservation and Management of Whales* (Seattle: University of Washington Press; and London: Butterworths, 1980), 12–14.

2. For detailed discussions of the economic circumstances of whaling in the postwar period, see J. N. Tønnessen and A. O. Johnsen, *The History of Modern Whaling* (Berkeley: University of California Press, 1982), chapters 28–31; and G. L. Small, *The Blue Whale* (New York: Columbia University Press, 1971), especially chapter 1.

3. See the introduction for a discussion and definitions of criteria I and II.

4. During the 1930s, in response to the huge Antarctic catches being made during this period and the effect this had on the price of whale oil, several international agreements intended to regulate the whaling industry were reached. These included the 1931 Geneva Convention for the Regulation of Whaling (several whaling countries including Japan, Germany, and the Soviet Union did not sign) and an agreement between the whaling companies in 1932 to limit whale oil production. This led to the adoption of the Blue Whale Unit (BWU), a bilateral agreement between Norway and The United Kingdom to limit their national industries in 1936 (several additional countries agreed to these limits in 1937 at a conference created by Norway) and the signing of further protocols on the original convention in 1938 and 1939. See J. E. Scarff, "The International Management of Whales, Dolphins, and Porpoises: An Interdisciplinary Assessment (Part One)," *Ecology Law Quarterly*, vol. 6, no. 323 (1977): 349–51; R. Gambell, "The IWC and the Contemporary Whaling Debate," in J. R. Twiss Jr. and R. R. Reeves, eds., *Conservation and Management of Marine Mammals*, 181 (Washington, DC: Smithsonian Institution Press, 1999); Tønnessen and Johnsen, *History of Modern Whaling*, chapters 24–26; and D. H. Cushing, *The Provident Sea* (Cambridge: Cambridge University Press, 1988), 155–56.

5. See Scarff, "International Management of Whales, Dolphins, and Porpoises," 349–51; and Tønnessen and Johnsen, *History of Modern Whaling*, chapters 24–26.

6. Tønnessen and Johnsen, *History of Modern Whaling*, 512–20.

7. Ibid., 492.

8. Allen, *Conservation and Management of Whales*, 12.

9. By the beginning of the twentieth century the whaling industry looked as though it would die a natural death due to the growing scarcity of whales and also declining demand for whale oil, caused by the discovery of mineral oil substitutes. Two important factors, however, allowed whaling to survive: the opening of the Antarctic made possible by the new ship-building technology and hunting methods, and the invention of a practical hydrogenation process, which made whale oil—previously used only for lighting and industrial purposes due to its "fluid form, poor quality and objectionable smell and taste"—suitable as a fat for producing margarine. In 1900, the world production of

whale oil stood at only 87,300 barrels. But by 1910, the year the hydrogenation process reached mass-production capability, world production had increased to 316,300 barrels in response to the burgeoning global demand for edible fats, particularly margarine. See Tønnessen and Johnsen, *History of Modern Whaling*, 227–36.

10. The 1931 Geneva Convention, which was not widely observed, protected right whales and females with calves in addition to requiring the licensing of whaling vessels and the collection of statistics. The 1937 agreement (also known as the First International Whaling Convention) had more support from whaling nations than the 1931 agreement and added grey whales to the list of protected whales, while also establishing the first "defined" Antarctic season. This agreement, which lasted for only one year, was renewed in 1938 with the added provisions of protection for Southern Hemisphere humpback whales (though only from pelagic and not coastal whaling) and the establishment of the Antarctic sanctuary area, which remained in place until the 1950s. See Allen, *Conservation and Management of Whales*, 23–24; and Tønnessen and Johnsen, *History of Modern Whaling*, especially chapters 24–26.

11. Tønnessen and Johnsen, *History of Modern Whaling*, 513–18.

12. Ibid., 526–29.

13. The BWU was adopted for the first time in the 1932 production agreement between whaling companies, which was intended to avoid a reoccurrence of the 1931 drop in prices caused by overproduction of whale oil. The blue whale had the highest yield and therefore became the standard against which other whales were measured (i.e., one blue whale unit = one blue whale = 2 fin whales = 2.5 humpback = 6 sei whales). Shore stations and floating factories were restricted to quotas (the first time quotas on production and whales killed were introduced) based on the number of barrels that must be extracted from each BWU taken. The intention was to reduce the 1930–31 season's barrel total by two-thirds or more and also to make more efficient use of each whale. Thus, the number of BWU was arrived at by dividing the number of barrels desired by the number of barrels required from each whale (110 in 1932 and then 115 in 1933). This formula resulted in an overall quota of 18,584 BWU for the 1932–33 season and 17,074 for the following season. See Tønnessen and Johnsen, *History of Modern Whaling*, 402–5.

14. See the schedule of the inaugural 1949 IWC meeting in IWC, *First Report*, 1950, 15.

15. Small, *Blue Whale*, 91–93.

16. Although some discussions of whaling during this time have given the impression that whaling ceased during the Second World War, the United King-

dom and Norway did in fact continue pelagic hunts in the Antarctic and elsewhere during the war but at greatly reduced levels. According to Tønnessen and Johnsen, "During the four seasons 1941–45 a total of some 7,000 whales were killed in the Antarctic, less than a fifth of the last peace-time season, 1938–39." While whaling continued "without serious restriction" off the coasts of South Africa and Japan, "the two largest whaling grounds outside the Antarctic" *History of Modern Whaling*, 472–77.

17. Tønnessen and Johnsen, *History of Modern Whaling*, 529–34; and W. G. Beasley, *The Modern History of Japan* (London: Weidenfeld and Nicolson, 1981), 291.

18. Nineteen countries attended the 1946 Washington whaling conference, with fourteen sending delegations and five sending observers. Of the seventy-two people in attendance, twenty-nine came from the United States (nine), the United Kingdom (twelve), and Norway (eight), representing the largest delegations at the conference. See Tønnessen and Johnsen, *History of Modern Whaling*, 499; and Scarff, "International Management of Whales, Dolphins, and Porpoises," 352–53.

19. The 1944 London Conference was the first agreement where the 16,000 BWU quota was mentioned. The 1945 meeting was better attended than the previous year's meeting, with forty-six delegates from twelve countries and was focused primarily on the issue of the European fats crisis and the creation of a control system for whaling capable of dealing with the shortage. See Tønnessen and Johnsen, *History of Modern Whaling*, 490 and 494; also P. Birnie, *International Regulation of Whaling: From Conservation of Whaling to Conservation of Whales and Regulation of Whale Watching*, vol. 1 (New York: Oceana Publications, 1985), 131–41.

20. See IWC, *First Report*, 1950, appendix 1, for the full text of the 1946 International Convention for the Regulation of Whaling.

21. Another whaling operation that remained undaunted by Norway's efforts was that of Aristotle Onassis, whose ships were registered under the flags of Panama and Honduras and crewed by Germans and Norwegians. Onassis's whaling expeditions, which continued until 1954 and were widely regarded as illegal, pirate operations, significantly increased the competition for the 16,000 BWU and also were unrestrained by the IWC's attempts at regulation. See Tønnessen and Johnsen, *History of Modern Whaling*, 534–38.

22. Ibid., 521–25.

23. Allen, *Conservation and Management of Whales*, 15.

24. Tønnessen and Johnsen, *History of Modern Whaling*, 521

25. Japan's whaling industry, unlike any of its competitors, benefited from both government financial assistance and exclusive access to a strong domestic market for whale meat. See Small, *Blue Whale*, 100–103 and 154–55.

26. Australia, Canada, France, Iceland, the Netherlands, Norway, Panama, South Africa, Sweden, the United Kingdom, the United States, and the Soviet Union. IWC, *First Report*, 1950, 3.

27. Ibid., appendix 1, 10.

28. Ibid., 9.

29. Ibid., Article III, 10.

30. Ibid., Article V, 11.

31. Ibid., 11–12.

32. See S. Holt, "The Whaling Game," *The Siren*, no. 35 (December 1987): 3–6.

33. S. Andresen, "Science and Politics in the International Management of Whales," *Marine Policy*, vol. 13, no. 2 (April 1989): 102.

34. According to the notion of "integrative bargaining and the veil of uncertainty," which is often cited in the regime theory literature, uncertainty, flexibility, and ambiguity in negotiations make it easier for parties to cooperate on problem issues by promoting "integrative or productive bargaining rather than distributive or positional bargaining." Oran R. Young and Gail Osherenko argue, "The veil of uncertainty refers to all those factors that make it difficult for individual participants to foresee how the operation of institutional arrangements will affect their interests over time. Individual parties' inability to predict a regime's impact on their welfare increases incentives to formulate provisions that are fair and equitable, which raises the probability that the parties can come up with institutional arrangements that are acceptable to all." Thus, "those engaged in regime formation may . . . act to thicken the veil of uncertainty by lengthening the time that the regime is expected to remain operative, enlarging the set of issues at stake, changing the membership of the group . . . , or devising ambiguous provisions open to later interpretation." See Young and Osherenko, "The Formation of International Regimes: Hypotheses and Cases," in Young and Osherenko, eds., *Polar Politics: Creating International Environmental Regimes*, 13 (Ithaca: Cornell University Press, 1993). For similar modes of bargaining concerning the Convention on International Trade in Endangered Species, see also P. Birnie, "International Environmental Law," in A. Hurrell and B. Kingsbury, eds., *The International Politics of the Environment* (Oxford: Oxford University Press, 1992), 52–54; and T. Brenton, *The Greening of Machiavelli* (London: Earthscan Publications, 1994), 102.

35. R. Gambell, "The Management of Whales and Whaling: An Historical Review of the Regulation of Commercial and Aboriginal Subsistence Whaling," *Arctic*, vol. 46, no. 2 (1993): 100.

36. Tønnessen and Johnsen, *History of Modern Whaling*, 511.

37. IWC, *First Report*, 1950, 4 and 6.

38. IWC, *Third Report*, 1952, 4.

39. According to Gambell, "The Scientific Committee meets during the two weeks before the Commission meeting; it may also hold special meetings during the year to consider particular topics. The information and advice it provides on the status of whale stocks form the basis on which the Commission develops whaling regulations. The Technical Committee is effectively a meeting of the entire Commission during which major items can go through a first round of discussion and, if necessary, be recommended to the Commission by simple majority. Gambell, "The IWC and the Contemporary Whaling Debate," 182.

40. See Birnie, *International Regulation of Whaling*, 177–79, for a succinct overview of the SC's organization up to the mid-1980s.

41. Recent changes to the rules of procedure for the IWC and the SC can be found in "Revised Rules of Procedure of the Scientific Committee," Annex M, Report of the Scientific Committee, *Journal of Cetacean Research and Management*, vol. 1 (1999): 247–50; and in "Revised Rules of Procedure to Incorporate Changes Agreed at IWC/53," Circular Communication to Commissioners and Contracting Governments IWC.CCG.209, IWC document NJG/JAC/29093, October 9, 2001.

42. T. Schweder, *Intransigence, Incompetence or Political Expediency? Dutch Scientists in the International Whaling Commission in the 1950s: Injection of Uncertainty*, IWC Scientific Committee Paper SC/44/o 13, 1992, 6.

43. See IWC, *Seventh Report*, 1956, 18, for delaying of and eventual withdrawal of Scientific Committee request for conference funding in 1954; and also IWC, *Second Report*, 1951, 6, for the commission's rejection of funding request for publication of SC reports. See Allen, *Conservation and Management of Whales*, 26, for irregular IWC meetings in the 1950s.

44. IWC, *First Report*, 1950, 11.

45. IWC, *Sixth Report*, 1955, 18–19.

46. See IWC, *Seventh Report*, 1956, 23–24; and Birnie, *International Regulation of Whaling*, 227–28.

47. Tønnessen and Johnsen, *History of Modern Whaling*, 565.

48. Cushing, *Provident Sea*, 162.

49. Colin W. Clark and Roland Lamberson's study suggests that there can

be utility in rapidly exhausting a natural resource such as whale stocks if the interest rate on an alternative investment is greater than the potential return offered by the net growth rate of the population. See Clark and Lamberson, "An Economic History and Analysis of Pelagic Whaling," *Marine Policy*, vol. 6, no. 2 (April 1982): 103–20.

50. Tønnessen and Johnsen, *History of Modern Whaling*, 487–88.

51. Ibid., 491.

52. IWC, *First Report*, appendix 3, 22.

53. T. Schweder, "Distortion of Uncertainty in Science: Antarctic Fin Whales in the 1950s," *Journal of International Wildlife Law and Policy*, vol. 3, no. 1 (2000): 78.

54. G. Elliot, *Whaling 1937–1967: The International Control of Whale Stocks* (Sharon, MA: Kendall Whaling Museum Monograph Series No. 10, 1997), 8.

55. The ban on pelagic hunting of humpbacks was lifted in 1949. At this time the humpback became the first species to be managed by a species-specific quota, which was set at 1,250 at the 1949 meeting. The rationale behind the lifting of the ban, which allowed humpbacks to be hunted in the Antarctic for the first time in a decade, was to relieve pressure on the blue whale stocks. Australia, however, where humpbacks were hunted from land stations on the eastern and western coasts, opposed the move. IWC, *First Report*, 1950, 28.

56. See the first (1950), second (1951), and third (1952) IWC reports.

57. Birnie, *International Regulation of Whaling*, 217.

58. Ibid.

59. IWC, *Fourth Report*, 1952, 8; and Birnie, *International Regulation of Whaling*, 219–20

60. Allen, *Conservation and Management of Whales*, 64–66.

61. "Report of the Scientific Committee," IWC fifth meeting (1953), Document XVII, 1–3.

62. Schweder, "Distortion of Uncertainty in Science," 78; and Birnie, *International Regulation of Whaling*, 221.

63. "The Condition of the Antarctic Whale Stocks," appendix to the Report of the Special Scientific Sub-Committee, IWC fifth meeting (1953), Document II.

64. "Memorandum of the Dutch Delegation on the Reduction of the 16,000 Blue Whale Units to the IWC Scientific Committee," IWC fifth meeting (1953), 1–3.

65. Ibid. See also Schweder, "Distortion of Uncertainty in Science," 79.

66. See IWC, *Fifth Report*, 1954, 3; and Birnie, *International Regulation of Whaling*, 221.

67. "Memorandum of the Dutch Delegation," 3.
68. IWC, *Sixth Report*, 1955, 20.
69. Tønnessen and Johnsen, *History of Modern Whaling*, 549–52; and IWC, *First Report*, 1950, 28.
70. IWC, *Sixth Report*, 1955, 15.
71. Ibid., 19.
72. Ibid., 19–20.
73. Ibid., 20.
74. Ibid.
75. IWC, *Seventh Report*, 1956, 23–24.
76. Schweder, "Distortion of Uncertainty in Science," 80. See also Birnie, *International Regulation of Whaling*, 228.
77. IWC, *Seventh Report*, 1956, 6.
78. Ibid.; and *Ninth Report*, 1958, 17.
79. IWC, *Seventh Report*, 1956, 6. The situation here is unclear in that the IWC annual report and other sources report only seven objections (see IWC, *Eighth Report*, 1957, 12) and do not include the original Dutch objection quoted here (which makes a total of eight). It appears highly unlikely that the Dutch commissioner would not have maintained an objection to the quota reduction and that the other governments would have done so alone, given the Dutch propensity for opposing quota reductions and also the fact that the Dutch commissioner initially voted against the reduction while some of the objecting governments originally voted for it.
80. IWC, *Sixth Report*, 1955, 20.
81. Ibid.
82. UK commissioner Mr. Wall speaking at the Technical Committee meeting during the IWC's 1955 meeting in Moscow, quoted in Birnie, *International Regulation of Whaling*, 229.
83. Tønnessen and Johnsen, *History of Modern Whaling*, 575.
84. Ibid., 564.
85. IWC, *Eighth Report*, 1957, 12.
86. Ibid.
87. Ibid., 19.
88. Ibid.
89. IWC, *Ninth Report*, 1958, 3.
90. Ibid.
91. IWC, *Tenth Report*, 1959, 3.
92. Schweder, "Distortion of Uncertainty in Science," 88 (personal correspondence by the author with Laws).

93. Tønnessen and Johnsen, *History of Modern Whaling*, 585.
94. IWC, *First Report*, 1950, 11.
95. Elliot, *Whaling 1937–1967*, 11.

3 / THE ANTARCTIC COLLAPSE

1. IWC, *Tenth Report*, 1959, 5.
2. See the introduction for a discussion and definitions of criterion I and criterion II.
3. See chapter 5 for a discussion of the IWC's current interpretation and application of the precautionary principle.
4. IWC, *Thirteenth Report*, 1962, 29–30 and 45–48; IWC, *Sixteenth Report*, 1966, 26–38.
5. IWC, *Tenth Report*, 1959, pp. 5-6.
6. See J. N. Tønnessen and A. O. Johnsen, *The History of Modern Whaling* (Berkeley: University of California Press, 1982), chapter 32, for a detailed account of the national quotas negotiations.
7. IWC, *Tenth Report*, 1959, 3 and 24.
8. Tønnessen and Johnsen, *History of Modern Whaling*, 589.
9. Ibid., 591.
10. IWC, *Tenth Report*, 1959, 24.
11. Ibid., 25–26.
12. Tønnessen and Johnsen, *History of Modern Whaling*, 589–91.
13. G. Elliot, *Whaling 1937–1967: The International Control of Whale Stocks* (Sharon, MA: Kendall Whaling Museum Monograph Series No. 10, 1997), 11 (my italics). Several other authors also have argued that the Dutch demands for a disproportionately high quota for their one expedition were a major cause of the national quotas talks failing to produce an agreement before the close of the 1959 meeting. Tønnessen and Johnsen, Small, Schweder, and Scarff, like Elliot, all have identified the Dutch demands for 1,200 BWU as the main obstacle to national quotas being set at the time. Birnie, however, also attributes responsibility to the Japanese and the Norwegians, which, in the case of Norway, may be a little unfair given the various concessions offered by the Norwegians in the hope of securing an agreement. See G. L. Small, *The Blue Whale* (New York: Columbia University Press, 1971), 193–94; P. Birnie, *International Regulation of Whaling: From Conservation of Whaling to Conservation of Whales and Regulation of Whale Watching*, vol. 1 (New York: Oceana Publications, 1985), 248–54; Tønnessen and Johnsen, *History of Modern Whal-*

ing, 589–94; T. Schweder, "Distortion of Uncertainty in Science: Antarctic Fin Whales in the 1950s," *Journal of International Wildlife, Law and Policy*, vol. 3, no. 1 (2001): 83; and J. E. Scarff, "The International Management of Whales, Dolphins and Porpoises: An Interdisciplinary Assessment (Part One)," *Ecology Law Quarterly*, vol. 6, no. 323 (1977): 361.

14. Tønnessen and Johnsen, *History of Modern Whaling*, 593–94.

15. IWC, *Ninth Report*, 1958, 3; and *Tenth Report*, 1959, 3.

16. See page 60 in chapter 2 for Commissioner Wall's comments at the Technical Committee meeting during the IWC's 1955 annual meeting in Moscow.

17. IWC, *Tenth Report*, 1959, 16.

18. IWC, *Eleventh Report*, 1960, 17; and Birnie, *International Regulation of Whaling*, 248.

19. Drion disputed the reliability of evidence that indicated the age distribution within certain stocks of fin whales had changed (i.e., considerably fewer older whales in the catch), suggesting a "rapid deterioration of the stock." Drion's opinion was that "the evidence previously put forward was insufficient for a recommendation to the Commission in regard to the total catch, and that the new figures from the Sanctuary area did not necessarily indicate a real decline in the local stock." IWC, *Ninth Report*, 1958, 22.

20. Tønnessen and Johnsen, *History of Modern Whaling*, 595 (my italics).

21. IWC, *Eleventh Report*, 1960, 16.

22. Ibid., 6.

23. Ibid., 17.

24. Tønnessen and Johnsen, *History of Modern Whaling*, 595.

25. Ray Gambell, interview by author, London, July 2001.

26. IWC, *Twelfth Report*, 1961, 15.

27. According to Tønnessen and Johnsen, this decision was bitterly opposed by the Norwegian whaling companies because Norway had rejoined without any progress on national quotas being made. The Norwegian companies were so incensed that they removed all representation from the State Whaling Council and refused to take any further part in future quota negotiations. The government explained its decision by saying it needed to take Norway's other foreign policy interests into account, including the merchant navy, fishing limit agreements, and the government's image concerning its commitment to international agreements. See Tønnessen and Johnsen, *History of Modern Whaling*, 602.

28. IWC, *Twelfth Report*, 1961, 3.

29. Ibid., 3–4 and 16.

30. Ibid., 4.

31. Ibid.
32. Ibid., 16.
33. Ibid.
34. Birnie, *International Regulation of Whaling*, 257–58.
35. IWC, *Eleventh Report*, 1960, 23.
36. Ibid., 25–26.
37. IWC, *Twelfth Report*, 1961, 14.
38. Ibid., 16.
39. Ibid.
40. Ibid., 8 and 16.
41. IWC, *Thirteenth Report*, 1962, 36–38; and IWC, *Twelfth Report*, 1961, 23 (for the last quote).
42. IWC, *Twelfth Report*, 1961, 20.
43. IWC, *Thirteenth Report*, 1962, 37.
44. Ibid., 39–40
45. Ibid., 40–41.
46. Ibid., 40.
47. Ibid., 37.
48. Ibid., 19.
49. Given the crisis the IWC found itself in and its members' apparent unwillingness to provide additional funding in order to deal with the question of sustainable quotas, it is difficult not to at least suspect that many in the commission were only paying lip service to the goal of achieving sustainable long-term management of the industry and the animals it depended upon. As Small observed, "It is both tragic and ironic that the excuse for inactivity by the Commission in 1962 was the absence of a committee report that was not prepared for lack of funds. The cost of the committee's work was £8,000—the market value equivalent of 4 blue whales." See Small, *Blue Whale*, 199.
50. Ibid., 22–23.
51. IWC, *Fourteenth Report*, 1964, 20–21.
52. IWC, *Thirteenth Report*, 1962, 8.
53. Ibid., 29–30.
54. Ibid., 17.
55. Norway had first suggested the idea of national quotas in 1955. But from the time of the scheme's formal proposal in the IWC by the United Kingdom in 1958 until its eventual conclusion in 1962, the Antarctic nations had required a total of eleven international conferences and the better part of four of the IWC's meetings to come up with an agreement that looked remarkably

similar to a proposal made by Norway in 1958 when negotiations had first begun. Norway's 1958 proposal, made before the United Kingdom and Norway's sale of factory ships to Japan, offered the following shares of the 15,000 BWU Antarctic quota (with the BWU equivalent in parentheses): Norway 33.63 percent (5,045); Japan 26.37 percent (3,955); the United Kingdom 15 percent (2,250); the Netherlands 5 percent (750); the Soviet Union 20 percent (3,000). See Tønnessen and Johnsen, *History of Modern Whaling*, 591; and IWC, *Tenth Report*, 1959, 5–6 and 24–25.

56. Ibid., 602–3; and IWC, *Twelfth Report*, 1961, 4.
57. Tønnessen and Johnsen, *History of Modern Whaling*, 603.
58. IWC, *Thirteenth Report*, 1962, 17 and 33–36.
59. Ibid., 3; and Tønnessen and Johnsen, *History of Modern Whaling*, 604–5.
60. IWC, *Thirteenth Report*, 1962, 4–5.
61. Tønnessen and Johnsen, *History of Modern Whaling*, 605.
62. IWC, *Fourteenth Report*, 1964, 14.
63. Ibid., 15.
64. Ibid.
65. Ibid., 16.
66. Ibid.
67. Ibid., 17.
68. Ibid., 5.
69. Ibid., 14.
70. Ibid.
71. IWC, *Seventh Report*, 1956, 16.
72. IWC, *Fourteenth Report*, 1964, 19–20.
73. Ibid., 35.
74. Ibid.
75. IWC, *Sixth Report*, 1955, 19–20.
76. IWC, *Eighth Report*, 1957, 23–24. In 1956, Ottestad developed a model independent of age determination methods, with assumed mortality and recruitment rates (i.e., his method was independent of age determination estimates), that supported Laws and Ruud's findings that fin stocks were declining, although the parameters of Ottestad's model differed from later (and better) estimates. See D. H. Cushing, *The Provident Sea* (Cambridge: Cambridge University Press, 1988), 158.
77. IWC, *Fourteenth Report*, 1964, 41-67.
78. T. D. Smith, *Scaling Fisheries: The Science of Measuring the Effects of Fishing, 1855–1955* (Cambridge: Cambridge University Press, 1994), 263–66.
79. IWC, *Fourteenth Report*, 1964, 36–37 (my italics in final paragraph).

See also Cushing, *Provident Sea*, 157–62, for another useful and easily followed summary of the COT's methods and assessment goals.

80. IWC, *Fourteenth Report*, 1964, 38.
81. Ibid., 40.
82. Ibid.
83. Ibid., 94.
84. Ibid., 25.
85. IWC, *Sixth Report*, 1955, 18–19.
86. IWC, *Fourteenth Report*, 1964, 37 and 41.
87. Ibid., 3–4.
88. IWC, *Thirteenth Report*, 1962, 4; IWC, *Fourteenth Report*, 1964, 3–4. See also Tønnessen and Johnsen, *History of Modern Whaling*, 603–7.
89. IWC, *Fourteenth Report*, 1964, 4.
90. IWC, *Fifteenth Report*, 1965, 17. See also Birnie, *International Regulation of Whaling*, 320–21.
91. IWC, *Fifteenth Report*, 1965, 17.
92. Ibid., 16–17.
93. Ibid., 4; and IWC, *Thirteenth Report*, 1962, 4.
94. IWC, *Fourteenth Report*, 1964, 82.
95. For a discussion of the pygmy blue whale and its differences with the larger "true blue whale," see S. L. Perry, D. P. DeMaster, and G. K. Silber, "Special Issue: The Great Whales: History and Status of Six Species Listed as Endangered Under the U.S. Endangered Species Act of 1973," *Marine Fisheries Review*, vol. 61, no. 1 (1999), http//spo.nwr.noaa.gov/mfr611/mfr611.htm.
96. IWC, *Twelfth Report*, 1961, 29–30; and IWC, *Fourteenth Report*, 1964, 82.
97. IWC, *Fifteenth Report*, 1965, 17.
98. Ibid.
99. Ibid., 18.
100. Small, *Blue Whale*, 200–202.
101. IWC, *Fourteenth Report*, 1964, 82.
102. Ibid., 94–95.
103. IWC, *Fifteenth Report*, 1965, 49.
104. Ibid.
105. IWC, *Fifteenth Report*, 1965, 32.
106. Birnie, *International Regulation of Whaling*, 257–58.
107. IWC, *Fifteenth Report*, 1965, 3.
108. Ibid.
109. Tønnessen and Johnsen, *History of Modern Whaling*, 622.

4 / THE WORM TURNS

1. IWC, *Twelfth Report*, 1961, 16.
2. "Verbatim Report of the Plenary Sessions," IWC sixteenth meeting (1964), IWC/16/15, p. 25.
3. P. Hammond, *The Conservation and Management of Small Cetaceans: A Review of the Options*, IWC Scientific Paper SC/43/SM12, p. 1.
4. At the 1963 meeting, just prior to the vote on the Japanese proposal for 10,000 BWU as the 1963–64 Antarctic quota, U.S. commissioner Kellogg proposed that the commission should reaffirm its 1960 resolution "that the Commission should not later than 31st of July, 1964 bring the Antarctic catch limit into line with the scientific findings having regard to the provisions of paragraph 2 of Article V of the Convention." The resolution was passed by ten votes with Japan, the Netherlands, and the Soviet Union reserving their position on any obligations concerning the 1960 resolution and its impact on the 1965–66 season's catch limit. See the "Verbatim Report of the Plenary Sessions," IWC fifteenth meeting (1963), IWC/15/17, pp. 73–76.
5. IWC, *Fifteenth Report*, 1965, 3.
6. Ibid., 17.
7. Ibid., 47. But, as Gambell has observed, "It [is] worth pointing out that we now know, because of recent information on the falsification of the Soviet catch records up to about 1972 (when the International Observer Scheme came into effect), that the agreement between the catch predicted by the Committee of Three and the catch recorded for 1963/64 was even more fortuitous." E-mail correspondence with Ray Gambell, August 13, 2002.

The full extent of the Soviet Union's duplicity was not revealed until after the USSR's collapse. At the IWC's 1994 meeting, Russian scientists revealed to the commission that the Soviet fleets had falsely reported catch data involving tens of thousands of blue, pygmy blue, fin, sei, humpback, and Bryde's whales, in the Antarctic and elsewhere, that were taken in addition to the numbers the Soviet Union had transmitted to the Bureau of International Whaling Statistics. See "Chairman's Report," appendix 6, IWC *Forty-fifth Report*, 1995; "Report of the Scientific Committee," in particular "Catch History Revisions" in Annex E, Report of the Sub-Committee on Southern Hemisphere Baleen Whales, also in IWC, *Forty-fifth Report*, 1995; and IWC, *Forty-sixth Report*, 1996, 131–38.
8. IWC, *Fifteenth Report*, 1965, 48.
9. Ibid.
10. Ibid.

11. IWC, *Fourteenth Report*, 1964, 95.

12. IWC, *Fifteenth Report*, 1965, 48-49.

13. Ibid., 33.

14. Ibid., 30–33 and 47–49.

15. "Verbatim Report of the Plenary Sessions," IWC sixteenth meeting (1964), IWC/16/15, p. 34.

16. Ibid., 72.

17. Ibid., 74.

18. IWC, *Fifteenth Report*, 1965, 32-33.

19. IWC, *Sixteenth Report*, 1966, 17–18; and "Verbatim Report of the Plenary Sessions," IWC sixteenth meeting (1964), IWC/16/15, pp. 76–77.

20. IWC, *Sixteenth Report*, 1966, 17.

21. Ibid., 18.

22. IWC, *Fifteenth Report*, 1965, 49.

23. Ibid., 31.

24. Ibid., 31–32.

25. Ibid., 32 (my italics).

26. "Verbatim Report of the Plenary Sessions," IWC sixteenth meeting (1964), IWC/16/15, pp. 57–60.

27. After the 1964–65 season, the number of Antarctic pelagic fleets dropped from 15 to 10, with Japan going from 7 to 5 and Norway and the Soviet Union each dropping from 4 to 2. By the 1967–68 season, Norway had only 1 fleet in operation as compared with 4 for Japan and 3 for the Soviet Union. This was the last Norwegian fleet to operate in the Antarctic. In the following season, Norway did not participate in any Antarctic hunting, but in the 1969–70 season the Norwegians unsuccessfully experimented with a combined factory/catcher that managed a total of only 6 BWU.

28. IWC, *Sixteenth Report*, 1966, 4.

29. Ibid., 3–4.

30. Ibid., 23.

31. IWC, *Fifteenth Report*, 1965, 23–26. The idea of an IOS was first floated by Norway at the 1955 meeting, but negotiations were slowed by the ICRW's lack of provision for such a scheme and also by the national quotas dispute. After the Antarctic nations finally agreed in 1963 on how the IOS should operate, its implementation then was delayed by Soviet demands for a renegotiation of the existing national quotas scheme (opposed by Japan, which now had the biggest share) and for the inclusion of land station catches in the Antarctic quota (opposed by the British, whose land station in South Georgia represented the United Kingdom's only remaining Antarctic interest after 1963).

32. IWC, *Sixteenth Report*, 1966, 24.

33. "Verbatim Report of the Plenary Sessions," IWC sixteenth meeting (1964), IWC/16/15, p. 69.

34. IWC, *Sixteenth Report*, 1966, 24 (my italics).

35. IWC, *Sixteenth Report*, 1966, 34.

36. Ibid., 55.

37. Ibid., 54. In the 1964–65 season the Soviet Union recorded only twenty pygmy blue catches while Japan recorded none.

38. IWC, *Seventeenth Report*, 1967, 19; IWC, *Eighteenth Report*, 1968, 18; and IWC, *Nineteenth Report*, 1968, 16–17.

39. Following the 1966 meeting, the Soviet Union, Japan, Norway, and the United Kingdom met to renegotiate the allotment of national quotas under the 3,500 BWU Antarctic limit. The new agreement gave Japan 1,633 BWU, the Soviet Union 1,067, and Norway 800. The practice of buying quota shares, as Japan earlier had done with the United Kingdom, the Netherlands, and Norway, also was halted by mutual agreement. See IWC, *Eighteenth Report*, 1968, 9–10.

40. Ibid., 54. The FAO Assessment Group, which included the COF scientists plus one FAO scientist, believed the sustainable yield for sei whales to be only 3,700 whereas the Japanese scientists believed the number could be as high as 7,000. There was, however, general agreement that some sei stocks were at their MSY level.

41. IWC, *Nineteenth Report*, 1968, 18.

42. IWC, *Nineteenth Report*, 1968, 29.

43. At a meeting in Japan shortly after the 1968 IWC meeting, the three remaining Antarctic nations agreed to reduce quotas (in BWU) as follows: Norway 731; Japan 1493; the Soviet Union 976. IWC, *Twentieth Report*, 1970, 9.

44. Ibid., 7.

45. By 1965, the price of whale oil per long ton (six barrels) had returned to the 1957 level of £85, but by 1968 the price fell by almost half to £46. In 1969, the price again began to rise, in accordance with the increasing shortage of supply, and peaked in 1974 at a record high of £537. In the 1973–74 season, however, the total baleen catch shared by Japan and the Soviet Union amounted to only 1,376 BWU and included more than 7,000 minke whales. See J. N. Tønnessen and A. O. Johnsen, *The History of Modern Whaling* (Berkeley: University of California Press, 1982), 750 and 753.

46. Ratios from IWC, *Twenty-first Report*, 1971, 18.

47. Tønnessen and Johnsen, *History of Modern Whaling*, 631–35.

48. IWC, *Twenty-second Report*, 1972, 7.

49. IWC, *Twentieth Report*, 1970, 34.

50. IWC, *Twenty-first Report*, 1971, 19.
51. IWC, *Twentieth Report*, 1970, 34.
52. Ibid., 35.
53. Ibid.
54. Ibid.
55. Ibid.
56. In the 1966–67 season, 2,893 fin whales and 12,368 sei whales were caught. By the 1968–69 season, with only the Japanese and Soviet fleets in operation, fin whale catches had increased to 3,014 while sei whale catches had plummeted to 5,770.
57. IWC, *Twenty-first Report*, 1971, 26 (my italics).

5 / SCIENTIFIC UNCERTAINTY AND THE EVOLUTION OF THE SUPERWHALE

1. Prior to the early 1960s, all IWC members represented active whaling countries, with the exception of Mexico. In the decade preceding the UN's Stockholm conference in 1972, the number of active whaling countries steadily declined until only eight of the IWC's fourteen members maintained commercial whaling industries. By the early 1980s, the commission was made up of only ten whaling members compared to thirty nonwhaling members. See R. Gambell, "The IWC and the Contemporary Whaling Debate," in J. R. Twiss and R. R. Reeves, eds., *Conservation and Management of Marine Mammals* (Washington, DC: Smithsonian Institution Press, 1999), 184.
2. IWC, *Twentieth Report*, 1970, 10; IWC, *Twenty-first Report*, 1971, 11.
3. IWC, *Twenty-fourth Report*, 1974, 24.
4. Ibid., 23–24.
5. Ibid., 24.
6. Other U.S. efforts to protect the larger whale species around this time included the Endangered Species Act of 1969, which banned all U.S. trade in whale-related products; the Marine Mammal Protection Act of 1972, which imposed a moratorium on all domestic hunting of marine mammals; and the 1972 Pelly Amendment to the Fisherman's Protective Act, which gave the president discretionary power to ban any and all fishing imports from states believed to be in violation of international agreements on the regulation and conservation of fisheries, including whaling. By the late 1970s, however, Congress was moving to further strengthen the Pelly Amendment with the introduction of the Packwood-Magnuson Amendment—an amendment that made the impo-

sition of sanctions automatic following "a finding by the Secretary of Commerce that a state is diminishing the effectiveness of the Whaling Convention or, under the Pelly Amendment, that a country is diminishing the effectiveness of any international fishery conservation program." See E. K. Flory, "Construing the Pelly and Packwood-Magnuson Amendments: The D.C. Circuit Court Harpoons Executive Discretion," *Washington Law Review*, vol. 61, no. 631 (1986): 631-36 (especially notes 19-26).

The legal interpretation and application of U.S. laws aimed at influencing the international exploitation of whale stocks and other fisheries, however, is extremely problematic due to ambiguities within the various legislation. A former attorney for the U.S. National Marine Fisheries Service who helped draft the Marine Mammal Protection Act has described it as "one of the most confusing and complex laws ever adopted by the United States. Many of its key provisions have language that is contradictory." See R. M. Parsons, *The US Marine Mammal Protection Act: A Report to the High North Alliance on the Waiver Process and Exemptions to the Embargo on Marine Mammal Products*, http://www.highnorth.no/Library/Trade/GATT_WTO/us-ma-ma.htm (report contracted by the High North Alliance in February 1996; released in November 1997).

7. IWC, *Twenty-fourth Report*, 1974, 24.

8. See note 1, this chapter.

9. At the 1971 IWC meeting, the SC informed the commission that it had been "unable to reach a single estimate for the sustainable yield [of the Antarctic fin whale stock] in 1971/72. Most members of the Committee believed that the best estimate was 2,200 whales (1,100 blue-whale-units), while the Japanese scientists estimated a yield of 4,250 (2,125 blue-whale units) with which the USSR scientists agreed." In London the following year, using new data from the 1971-72 season, the SC increased its sustainable yield estimate for the Antarctic fin stock to 3,200, but recommended that catches be held below this level. Proposals for quotas of 2,000 and then 1,800 were made in the plenary before a proposal by Japan, seconded by the Soviet Union, for 1,950 finally achieved the required three-quarters majority. See IWC, *Twenty-third Report*, 1973, 19; and IWC, *Twenty-fourth Report*, 1973, 28-29.

10. See chapter 4, note 31.

11. IWC, *Twenty-fourth Report*, 1974, 27.

12. IWC, *Twenty-third Report*, 1973, 22-23.

13. As was noted in chapter 4, note 7, the full extent of the Soviet Union's false reporting was not made clear to the IWC until its 1994 and 1995 meetings, although various infractions reports from the 1950s and 1960s demon-

strated that, at the very least, undersized whales were being taken. There had long been suspicions of misreporting in the commission and the whaling industry. Elliot, for example, writes, "From the earliest days of the IWC there were grave suspicions that the Russians, and to a lesser extent the Japanese, were breaking the Convention, taking undersized whales or protected species, fishing outside the season, and over-reporting their catch so that the season was shortened and they could be left to whale on their own after everyone else had gone home. The Russians were caught at this on several occasions and were reported to the IWC, which investigated the alleged infractions. Nothing, of course, could be done in the face of blank Russian denials, and other governments were in any case not keen to stir up a row which might push the USSR to walk out and resign from the Convention" (G. Elliot, *Whaling 1937–1967: The International Control of Whale Stocks* [Sharon, MA: Kendall Whaling Museum Monograph Series No. 10, 1997], 8–9).

14. IWC, *Twenty-third Report*, 1973, 38.

15. IWC, *Twenty-fourth Report*, 1974, 26–27. Whaling countries operating outside the IWC—in particular Brazil, Chile, Portugal, Peru, and Spain—were urged to adhere to the ICRW and its schedule as soon as possible. Concerning staffing, the IWC so far had been operating out of the United Kingdom's Ministry for Agriculture using only part-time staff. An ad hoc committee was appointed to consider plans for strengthening the secretariat, which in the committee's report the following year included "the establishment of an office for the Commission and the appointment of full time staff including a scientist as its chief officer." The Japanese and Soviet commissioners, however, expressed reservations over the provision of additional funding by members "in view of the uncertain future of whaling, particularly having regard to certain decisions reached at this meeting." See IWC, *Twenty-fifth Report*, 1975, 31.

16. IWC, *Twenty-fourth Report*, 1974, 69.

17. IWC, *Twenty-fifth Report*, 1975, 26.

18. The strong U.S. support for the moratorium was only one of a number of policy initiatives made by the Nixon administration concerning environmental issues—a strong indication, perhaps, of the extent to which governments at this time already were becoming aware of the growing political importance of environmental issues. As U.S. scientist and former IWC commissioner William Aron has noted, "The moratorium appears almost out of nowhere in the early seventies, but one has to put it into a much broader perspective. This was a time, and I think most people don't realize this, that our president, Richard Nixon, took major steps towards the environment. In the Nixon administration—almost at one time—we passed the National Environmental Policy Act; we created the

Environmental Protection Agency; we created the National Oceanic and Atmospheric Administration; we changed our water and air quality laws—all for the benefit of the environment. So, some of the biggest steps in the past forty and fifty years for the environment, came through the Nixon White House. Now, during that time period, there was—at least in the United States—a wide variety of environmental organizations, many you might expect [were] contrary in interest to one another's views. . . . At the same time there was a strong desire to build an environmental ethic. And the whaling issue was ready-made for pulling these organizations together." William Aron, interview by author, Tokyo, March 2001.

19. W. W. Fox, L. M. Talbot, W. Aron, and V. B. Scheffer, "Biological Rationale and Evidence for Ceasing to Take Antarctic Fin Whales," Scientific Committee document SC/25/33, in IWC, *Twenty-fourth Report*, 1974, 190–93.

20. IWC, *Twenty-fifth Report*, 1975, 26.

21. Ibid., 30.

22. The average catch per catcher day for fin whales by pelagic expeditions in the Antarctic had fallen from 0.32 in the 1971–72 season to 0.23 in the 1972–73 season. In the 1973–74 season, the figure dropped again to 0.19. See IWC, *Twenty-fourth Report*, 1974, 7; and IWC, *Twenty-fifth Report*, 1975, 8.

23. IWC, *Twenty-fifth Report*, 1975, 6–7.

24. IWC, *Twenty-fourth Report*, 1974, 42.

25. Ibid.

26. Ibid.

27. IWC, *Twenty-fifth Report*, 1975, 27.

28. Ibid., 27–28.

29. According to Aron, a member of the SC at the time who later became the U.S. commissioner, "it was expressly stated by one of the Japanese scientists that they could not accept the low number urged by most of the Scientific Committee members [i.e., an Antarctic quota of 5,000 minke whales] because that number was too low to meet their industry's need." W. Aron, "Science and the IWC," in R. L. Friedheim, ed., *Toward a Sustainable Whaling Regime* (Seattle: University of Washington Press, 2001), 111.

30. See IWC reports fifteen through thirty-three.

31. J. E. Scarff, "The International Management of Whales, Dolphins, and Porpoises: An Interdisciplinary Assessment (Part One)," *Ecology Law Quarterly*, vol. 6, no. 323 (1977): 369. See also *Newsweek*, "Song of the Whales," July 8, 1974.

32. In addition to the criticisms being made by the United States concerning the lack of data and various uncertainties over stock numbers, recruitment

rates, and so on, involved in the SC's estimates, a report presented to the IWC in 1974 by the FAO Marine Mammals Project (headed by Holt) and the Director General's Advisory Committee on Marine Resources Research (ACMMR) also pointed out a number of problems relating to the work of the SC. The report paid particular attention to the lack of important stock number and biological data and also the issue of how the resulting levels of uncertainty in the committee's advice should be interpreted and presented to the commission. See IWC *Twenty-fifth Report*, 1975, 253–60.

33. Ibid. See also W. Aron, W. Burke, and M. M. R. Freeman, "The Whaling Issue," *Marine Policy*, vol. 24 (2000): 180; W. Aron, W. Burke, and M. M. R. Freeman, "Flouting the Convention," *Atlantic Monthly*, May 1999, 22–23; and S. Andresen, "Science and Politics in the International Management of Whales," *Marine Policy* (April 1989): 107–9.

34. W. Aron, "Laws, Treaties and Evolution," lecture given at the symposium High Seas Fisheries and International Fishery Management Organization held by the Ministry of Foreign Affairs of Japan and the Association for the Comparative Study of Legal Cultures, March 9, 2001, Tokyo.

35. IWC, *Twenty-sixth Report*, 1976, 25.

36. Ibid. 26.

37. At the 1974 TC meeting, the United States had again proposed a moratorium on commercial hunting, seconded this time by Mexico. The proposal passed in the TC by a simple majority but again failed to gain the required majority in the plenary. The SC, after reviewing its statement in response to the moratorium at the 1973 IWC meeting, reported to the commission that "there was no biological requirement for the imposition of a blanket moratorium on all commercial whaling and agreed that the [previous year's] statement was still appropriate. It drew attention to the suggestion of possible competition between species whereby rebuilding of severely depleted stocks may not necessarily be maximized by a moratorium." Ibid., 25.

38. Ibid., 26.

39. G. P. Donovan, "The International Whaling Commission and the Revised Management Procedure," in E. Hallenstvedt and G. Blichfeldt, eds., *Additional Essays on Whales and Man* (Lofoten, Norway: High North Alliance, 1995), 5–6. See also IWC, *Twenty-seventh Report*, 1977, 6–8, for both the actual classification definitions and their application to the species under IWC management at the time as per the schedule amendments.

40. IWC, *Twenty-seventh Report*, 1977, 7–9.

41. Ibid., 6–9. See also Scarff, "The International Management of Whales, Dolphins, and Porpoises," 423–26. The 10 percent threshold below MSY was

referred to as Z percent, which was based on a discount rate of 4 percent that was intended to allow for natural mortality as a means of compensating for stock misclassification. At the time, Chapman argued for a wholly biologically based discount rate that did not take economic factors into account. He considered the latter would lead to populations being reduced to "well below their MSY levels" as per Clark's analysis of why the whaling industry traditionally opted for short-term maximization of profits over long-term sustainability. Thus, with 60 percent as the accepted MSY level for baleen stocks and Z percent equal to 10 percent, the protection level was set at 54 percent. For a discussion of the origins of the 54 percent level, see D. S. Butterworth and P. Best, "The Origins of the Choice of 54 Percent of Carrying Capacity as the Protection Level for Baleen Whale Stocks, and the Implications thereof for Management Procedures," SC/F92/Mg2 (Revised), in IWC, *Forty-fourth Report*, 1994, 491–97. For a discussion of why the low productivity of whale stocks led to overexploitation by industry, see C. W. Clark and R. Lamberson, "An Economic History and Analysis of Pelagic Whaling," *Marine Policy*, vol. 6, no. 2 (April 1982): 103–20.

42. Scarff, "The International Management of Whales, Dolphins, and Porpoises," 371; and IWC, *Twenty-seventh Report*, 1977, 23–25.

43. Ibid., 18.

44. Donovan, "The International Whaling Commission and the Revised Management Procedure," 5–6.

45. Aron, "Science and the IWC," 115–16.

46. At the 1982 IWC meeting, the Japanese commissioner pointed out that the SC "had stated in the past that there was no scientific justification for a blanket moratorium." Japan, Norway, Iceland, and South Korea all opposed the moratorium on account of its lack of a scientific basis as required by the ICRW. Switzerland, citing similar concerns, abstained from voting while several South American governments expressed concern regarding the moratorium's impact on the sovereignty of coastal states within their two-hundred-mile exclusive economic zones. See IWC, *Thirty-third* Report, 1983, 21.

47. The SC's 1979 report noted that, "Some members believe that in the event of a total moratorium the cessation of the flow of new data from commercial whaling operations, including sightings by scouting boats, CPUE indices, and biological samples, even if only for a limited period, would materially hinder the improvement of our understanding. Some members, however, doubted whether the omission of one or even a few years of additional data . . . derived from biological samples would have any detrimental consequences." IWC, *Thirtieth Report*, 1980, 47.

48. Ibid.

49. Sidney Holt, interview by author, London, July 2001.

50. See, for example, S. Holt, "Activities of the FAO/ACMMR Working Party on Marine Mammals," Annex X, Report of the Scientific Committee, IWC, *Twenty-fifth Report*, 1975, 253–60; S. Holt, "Criteria for Management—Weight or Numbers? The Relevance of Whaling Effort," SC/27/Doc 33 Revised, Report of the Scientific Committee, in IWC, *Twenty-sixth Report*, 1976, 409–13; S. Holt, "Whale Management Policy," SC/28/Doc 10, Report of the Scientific Committee, in IWC, *Twenty-seventh Report*, 1977, 133–37; S. Holt, "Some Implications of Maximum Sustainable Net Yield as a Management Objective for Whaling," SC/29/Doc 17, Report of the Scientific Committee, in IWC, *Twenty-eighth Report*, 1978, 191–93; S. Holt, "Proposal for a Modified Management Policy," SC/30/Doc 51, Report of the Scientific Committee, in IWC, *Twenty-ninth Report*, 1979, 327–33; and also D. Chapman, W. de la Mare, S. Holt, F. Pascal, "Moratorium Proposals," Annex M, Comments on Moratorium Proposals, Report of the Scientific Committee, in IWC, *Thirty-third Report*, 1983, 184.

51. E-mail correspondence with Ray Gambell, September 1, 2002. Aron also remembers Holt and Beddington playing major roles in the debates over uncertainty issues in the SC: "Both Holt and Beddington spent lots of time questioning virtually every analysis, Sidney, in particular, was always running new analyses and forcing reexaminations of issues that the rest of us thought had been long decided. This was a 'sand in the gears' approach that lengthened meetings, but produced no real changes." E-mail correspondence with William Aron, August 20, 2002.

52. IWC, *Thirtieth Report*, 1980, 39.

53. The Australian government decided to end all domestic whaling on the basis of the Australian Inquiry into Whales and Whaling led by Sir Sydney Frost. In its opening statement at the 1979 IWC meeting, Australia cited the report's findings as the basis for its support for an international ban on *all* whaling, including so-called aboriginal whaling. The Frost report based its recommendations on arguments concerning the "high intelligence" of whales, public opinion against whaling, and uncertainties in the NMP. See "Opening Statement by the Australian Commissioner," IWC thirty-first meeting (1979), London.

54. "Opening Statement of Richard A. Frank, U.S. Commissioner," IWC thirty-first meeting (1979), London. After the Carter administration's departure from office, Frank was forced to leave government and was then hired by the Japanese to represent them on whaling issues.

55. IWC, *Thirty-fifth Report*, 1985, 13.

56. IWC, *Thirty-third Report*, 1983, 20–21.

57. Ibid., 21.

58. IWC, *Thirty-fifth Report*, 1985, 13.

59. IWC, *Thirty-third Report*, 1983, 47.

60. For a statement by Allen and other scientists who opposed the moratorium's adoption, see K. R. Allen, D. Butterworth, P. Best, M. Cawthorn, M. Fraker, F. O. Kapel, E. Murphy, T. Øritsland, and C. J. Rørvik, "A Statement on Moratorium Proposals," Annex M, Comments on Moratorium Proposals, Report of the Scientific Committee, in IWC, *Thirty-third Report*, 1983, 183–84.

61. Gulland's own view on the moratorium was less than positive for two important reasons: (a) he believed the adoption of a blanket ban was similar to the BWU approach in that it failed to make any distinction between the circumstances of the different stocks; and (b) he also believed the moratorium had only further polarized the IWC and made the chances of cooperation more remote because it "convinced Japan and the USSR that many environmental groups were opposed to any use of living resources, however rational and controlled." See J. Gulland, "The End of Whaling?" *New Scientist*, October 29, 1988, 45.

62. IWC, *Thirty-third Report*, 1983, 40.

63. Andresen, "Science and Politics in the International Management of Whales," 111–12.

64. Ibid., 112.

65. IWC, *Thirty-third Report*, 1983, 40.

66. As Donovan has noted, the original wording of the moratorium only stated that "catch limits for the killing of whales for commercial purposes shall be zero," but before the vote was taken the extra wording including the comprehensive assessment and the "establishment of other catch limits" was added. The extra wording was added, in Donovan's view, "perhaps to indicate to the whaling countries that the proposal [i.e., the moratorium] seriously considered the possibility of whaling resuming." See G. P. Donovan, ed., preface to *The Comprehensive Assessment of Whale Stocks: The Early Years*, Special Issue 11 (Cambridge: IWC, 1989).

67. Ibid.

68. "Report of the Special Meeting of the Scientific Committee on Planning for a Comprehensive Assessment of Whale Stocks," in Donovan, *Comprehensive Assessment of Whale Stocks*, 3.

69. For an excellent discussion of the precautionary principle and also an overview of the various definitions and international declarations pertaining to it, see *Extracts and Summary from the Danish Environmental Protection Agency's Conference on the Precautionary Principle* (Copenhagen: Ministry of

Environment and Protection, Danish Environmental Protection Agency, May 29, 1998).

70. D. Bodansky "Scientific Uncertainty and the Precautionary Principle," *Environment*, vol. 33, no. 7 (September 1991): 5. For similar criticism of the precautionary principle see also D. Taverne, "Against Anti-Science," *Prospect*, December 1999; J. M. MacDonald, "Appreciating the Precautionary Principle as an Ethical Evolution in Ocean Management," *Ocean Development and International Law*, vol. 26 (1995): 255–86; and R. Bailey, "Precautionary Tale," *Reason*, April 1999.

71. Donovan, "The International Whaling Commission and the Revised Management Procedure," 7.

72. Ibid.

73. Pro-whaling governments, in particular Japan, Norway, and Iceland, have complained that agreement on the working details of the RMS, which includes inspection, international observation, and monitoring of catches and products, is being delayed by demands from governments opposed to whaling for very strict inspection and monitoring procedures that are, from an industry point of view, both impractical and also outside the terms of the ICRW. The pro-whaling governments contend that these demands are unjustified and are only a means of further delaying implementation of the RMP.

74. Donovan, "The International Whaling Commission and the Revised Management Procedure," 7.

75. See J. Cooke, "Simulation Studies of Two Whale Stock Management Procedures," in Donovan, *The Comprehensive Assessment of Whale Stocks*, 147–48; Donovan, "The International Whaling Commission and the Revised Management Procedure"; F. Kasamatsu, "Counting Whales in the Antarctic," in *Research on Whales*, 30–33 (Tokyo: Institute of Cetacean Research, 1995); and "Comprehensive Assessment Workshop on Catch Per Unit Effort (CPUE)," in Donovan, *The Comprehensive Assessment of Whale Stocks*, 15–19.

76. "Report of the Special Meeting of the Scientific Committee on Planning for a Comprehensive Assessment of Whale Stocks," 5.

77. Donovan, "The International Whaling Commission and the Revised Management Procedure," 7.

78. "Comprehensive Assessment Workshop on Management," in Donovan, *The Comprehensive Assessment of Whale Stocks*, 22.

79. D. S. Butterworth, "Sustainable Utilisation of Marine Mammal Resources," in Hallenstvedt and Blichfeldt, *Additional Essays on Whales and Man*, 12.

80. D. S. Butterworth, "Science and Sentimentality," *Nature*, vol. 357 (June 18, 1992): 533.

81. See IWC, *Forty-sixth Report*, 1996, 27 and 45–46.

82. S. Holt, *Minkes' Saga, Part II or Whatever Happened to Vesterfjord?* IFAW/IOI Technical Briefing, vol. 93, no. 7 (May 1993): 4.

83. Doug DeMaster, interview by author, Shimonoseki, Japan, May 2002.

84. Justin Cooke, interview by author, London, July 2001.

85. For a definition and some criticisms of the MSY concept see T. D. Smith, *Scaling Fisheries: The Science of Measuring the Effects of Fishing, 1885–1955* (Cambridge: Cambridge University Press, 1994), 264–66; and D. Ludwig, R. Hilborn, and C. Walters, "Uncertainty, Resource Exploitation, and Conservation: Lessons from History," *Science*, vol. 260 (April 2, 1993): 17–18.

86. Greg Donovan, interview by author, London, July 2001.

87. Institute of Cetacean Research, *Whale Management Under the International Whaling Commission's Revised Management Procedure (RMP)*, http://www.whalesci.org/research/management.html.

88. J. Lemons, "The Conservation of Biodiversity: Scientific Uncertainty and the Burden of Proof," in J. Lemons, ed., *Scientific Uncertainty and Environmental Problem Solving*, 221 (Oxford: Blackwell Science, 1996).

89. P. Harremoes, "Can Risk Analysis Be Applied in Connection with the Precautionary Principle?" in *Extracts and Summary from the Danish Environmental Protection Agency's Conference on the Precautionary Principle*, 30–35.

90. Lemons, "The Conservation of Biodiversity," 229.

91. Ibid (my italics).

92. E-mail correspondence with Greg Donovan, August 22, 2000.

93. "Opening Statement of U.S. Commissioner Richard A. Frank," IWC thirty-first meeting (1979), London.

94. Both the New Zealand and Australian opening statements to the 1979 meeting cited ethical concerns over the killing of whales, public opinion against whaling, and also problems with the NMP as the basis of their support for the moratorium. See "Opening Statement of New Zealand Commissioner B. J. Lynch," IWC thirty-first meeting (1979), London; and "Opening Statement of the Australian Commissioner," IWC thirty-first meeting (1979), London.

95. A. Kalland, "Whose Whale is That? Diverting the Commodity Path," in M. M. R. Freeman and U. P. Kreuter, eds., *Elephants and Whales: Resources for Whom?* 161 (Amsterdam: Gordon and Breach, 1994).

96. Prior to Spong's involvement in Greenpeace, he was dismissed from his research job at a Vancouver aquarium because he had announced to the media

that the orca he was working with had told him they wanted to be free. See M. Brown and J. May, *The Greenpeace Story* (Scarborough: Prentice-Hall Canada, 1989), 32–34.

97. E-mail correspondence with Paul Spong, April 22, 1999.

98. Ben White, interview by author, London, July 2001.

99. Kalland, "Whose Whale is That?" 163.

100. Ibid., 162–63.

101. In the early 1990s, a Gallup Poll on international attitudes to whales and whaling was conducted consisting of samples of five hundred adults from each of the following countries: Australia, Germany, Japan, Norway, England. The U.S. sample consisted of one thousand adults. The survey revealed that the general public in each country understood little about whales and whale hunting but also indicated that misconceptions were strongest among people from nonwhaling countries. For example, when asked if all large whales are currently in danger of extinction, 70 percent of Australians incorrectly answered "yes" as opposed to only 41 percent of Norwegians. When asked if it is true that some countries kill more than one thousand whales a year for scientific research, 70 percent of Australians, Germans, and British people incorrectly answered that it is true (67 percent in the United Stated) while only 40 percent of Japanese and 32 percent of Norwegians sampled answered incorrectly. See M. M. R Freeman and S. R. Kellert, "International Attitudes to Whales, Whaling and the Use of Whale Products: A Six Country Survey," in Freeman and Kreuter, *Elephants and Whales: Resources for Whom?* 298.

102. Kalland, "Whose Whale is That?" 163–64.

103. See L. Spencer, "The Not So Peaceful World of Greenpeace," *Forbes*, November 11, 1991; M. M. R. Freeman, "Science and Trans-Science in the Whaling Debate," in Freeman and Kreuter, *Elephants and Whales: Resources for Whom?* 143–57; A. Kalland, "Super Whale: The Use of Myths and Symbols in Environmentalism," in Blichfeldt (ed.), *11 Essays on Whales and Man*, 39–41 (Lofoten, Norway: High North Alliance, 1994); former Greenpeace International director Pete Wilkinson's commentary, "The World Needs More Than Protests," *Nature*, vol. 396 (December 10, 1998): 511–12; M. Shaw Bond, "Special Report: The Backlash Against NGOs," *Prospect*, April 2000; and advisor to the Institute of Cetacean Research Dan Goodman's letter in *Nature*, vol. 397 (January 28, 1999): 290.

104. H. Sorensen, "The Environment Movement and Minke Whaling," in Blichfeldt, *11 Essays on Whales and Man*, 28.

105. Ibid., 28–29.

106. Gulland, "The End of Whaling?" 45.

107. See Kalland, "Whose Whale is That?" 167; and Butterworth, "Science and Sentimentality," 532–34.

108. Kalland, "Whose Whale is That?" 167.

109. Butterworth, "Science and Sentimentality," 532.

6 / CONCLUSION

1. E-mail correspondence with William Aron, January 4, 2002.

2. P. M. Haas, "Introduction: Epistemic Communities and International Policy Coordination," in P. M. Haas, ed., *Knowledge, Power, and International Policy Coordination*, 14–15 (Columbia: University of South Carolina Press, 1997).

3. Haas likens epistemic communities to "Kuhn's broader sociological definition of a paradigm, which is 'an entire constellation of beliefs, values, techniques, and so on shared by members of a given community' and which governs 'not a subject matter but a group of practitioners.'" According to Haas, "what bonds members of an epistemic community is their shared belief or faith in the verity and applicability of particular forms of knowledge or specific truths." Ibid., 3.

4. Ibid., 23. Haas also suggests that while "knowledge is only accepted belief, not correct belief, correct beliefs may evolve over time." Ibid. The problem here, of course, is that even if "correct beliefs" were to evolve over time, how could we recognize them as such when (and if) they finally arrived?

5. Immediately following the commission's refusal to adopt the RMP in 1993, the SC's chairman, Philip Hammond, resigned in protest. In his letter of resignation, Hammond wrote, "what is the point of having a Scientific Committee if its unanimous recommendations on a matter of primary importance [i.e., the RMP] are treated with such contempt? . . . I have come to the conclusion that I can no longer justify to myself being the organizer of and the spokesman for a Committee whose work is held in such disregard by the body to which it is responsible." Philip Hammond, letter of resignation to the IWC, May 26, 1993, IWC Ref:IWC/2.1.

6. E-mail correspondence with Doug Butterworth, September 16, 2002.

7. M. J. Peterson, "Whalers, Cetologists, Environmentalists, and the International Management of Whaling," in Haas, *Knowledge, Power, and International Policy Coordination*, 147–86.

8. Ibid., 153–55.

9. Ibid., 149 and 185–86.

10. This recommendation to open the sanctuary was opposed by the Soviet commissioner. See IWC, *Sixth Report*, 1955, 19.

11. IWC, *Seventh Report*, 1956, 19.

12. E-mail correspondence with Peter Best, September 6, 2002.

13. See "Opening Statement by New Zealand," IWC forty-sixth meeting (1994), Puerto Vallarta.

14. As the policy positions of Australia and New Zealand demonstrate— along with even a short perusal of the literature produced by many anti-whaling NGOs—from the 1990s opposition to lifting the moratorium has increasingly been based on various ethical arguments against whaling. However, the moratorium's adoption was originally based on management uncertainties, and many anti-whaling governments still cite uncertainty over abundance estimates as a reason for the moratorium's continuation despite the commission's acceptance of the RMP.

Cooke's criticisms of the SC's Abundance Estimate Working Group's estimates for Northeast Atlantic minke stocks in 1996 (a controversy not resolved until 1998), and more recent claims by whaling opponents that the as yet incomplete reassessment of Southern Hemisphere minke stocks has revealed much smaller numbers than originally thought, are examples of the role that uncertainty issues still play in the commission. See the Chairman's Report, in IWC, *Forty-Seventh Report*, 1997, 30–32. For commentaries both supporting and opposing Cooke's position on the 1996 estimates, see S. Holt, "Coverage of the International Whaling Commission's 1996 Annual Meeting June 23–28, Aberdeen, Scotland," *Eco*, no. 3 (June 26, 1996); and *The International Harpoon*, no. 2 (June 25, 1996).

BIBLIOGRAPHY

Aldhous, P., and D. Swinbanks. "Whaling Ban Versus Science." *Nature*, vol. 351 (May 23, 1991): 259.
Allen, K. R. *Conservation and Management of Whales*. Seattle: University of Washington Press; London: Butterworths, 1980.
Andresen, S. "Science and Politics in the International Management of Whales." *Marine Policy* (April 1989): 99–117.
Aron, W. "Laws, Treaties and Evolution," Lecture given at the symposium High Seas Fisheries and International Fishery Management Organization held by the Ministry of Foreign Affairs of Japan and the Association for the Comparative Study of Legal Cultures, March 9, 2001, Tokyo.
———. "Save The Whalers," *The Wall Street Journal*, September 9, 1997.
———. "Science and the IWC." In R. L. Friedheim, ed., *Toward A Sustainable Whaling Regime*, 105–122. Seattle: University of Washington Press, 2001.
Aron, W., W. Burke, and M. M. R. Freeman. "Flouting the Convention." *Atlantic Monthly*, May 1999.
———. "The Whaling Issue." *Marine Policy*, vol. 24 (2000): 179–91.
Bailey, R. "Precautionary Tale," *Reason*, April 1999.
Bate, R., ed. *What Risk? Science, Politics and Public Health*. Oxford: Butterworth-Heinemann, 1999.
BBC Wildlife, "World Weary of Whaling Boors," July 1995.
Beasley, W. G. *The Modern History of Japan*. London: Weidenfeld and Nicolson, 1981.
Bernstein, P. L. *Against the Gods: The Remarkable Story of Risk*. New York: John Wiley and Sons, 1998).
Birnie, P. *International Regulation of Whaling: From Conservation of*

Whaling to Conservation of Whales and Regulation of Whale Watching. Vols. 1 and 2. New York: Oceana Publications, 1985.

———. "International Environmental Law." In A. Hurrell and B. Kingsbury, eds., *The International Politics of the Environment*, 51–84. Oxford: Oxford University Press, 1992.

Bodansky, D. "Scientific Uncertainty and the Precautionary Principle." *Environment*, vol. 33, no. 7 (September 1991): 4–5 and 43–44.

Brenton, T. *The Greening of Machiavelli.* London: Earthscan Publications, 1994.

Brown, M., and J. May. *The Greenpeace Story.* Scarborough: Prentice-Hall Canada, 1989.

Burns, J. H., and H. L. A. Hart, eds. *Collected Works of Jeremy Bentham: An Introduction to the Principles of Morals and Legislation.* London: The Athlone Press, 1970.

Butterworth, D. S. "Science and Sentimentality," *Nature*, vol. 357 (June 18, 1992): 532–34.

———. "Taking Stock: Science and Fisheries Management Entering the New Millennium." Lecture paper, University of Cape Town, August 1999.

Carr, E. H. *What Is History?* Oxford: Oxford University Press, 1961.

Chadwick, D. H. "Bottlenose Whales." *National Geographic*, vol. 194, no. 2 (August 1998): 79–89.

Chalmers, A. F. *What Is This Thing Called Science?* Saint Lucia: University of Queensland Press, 1999.

Clark, C. W., and R. Lamberson. "An Economic History and Analysis of Pelagic Whaling." *Marine Policy*, vol. 6, no. 2 (April 1982): 103–20.

Cole, C. F. "Can We Resolve Uncertainty in Marine Fisheries Management?" In Lemons, *Scientific Uncertainty and Environmental Problem Solving*, 233–63.

Crenson, M. "Particle Collider to Re-create Birth of Universe." *Daily Yomiuri*, March 28, 2000.

Cushing, D. H. *The Provident Sea.* Cambridge: Cambridge University Press, 1988.

Daily Yomiuri, "Australians Rally against Whaling." May 10, 1993.

Daily Yomiuri, "Issue of Whaling at a Standstill." May 10, 1993.

Daily Yomiuri, "Whales' Fate at Stake." March 25, 1993.

Derr, M. "To Whale or Not to Whale." *Atlantic Monthly*, October 1997.

Doyal, L., and I. Gough. *A Theory of Human Need.* London: Macmillan, 1991.

Drozdiak, W., and E. Pianin. "Allies Stunned by U.S. Kyoto Stance." *Daily Yomiuri*, March 30, 2001.

Dworkin, G. "Paternalism." In G. Dworkin, ed., *Mill's On Liberty: Critical Essays*, 61–82. Lanham, MD: Rowman and Littlefield Publishers, 1997.

Elliot, G. *Whaling 1937–1967: The International Control of Whale Stocks*. Sharon, MA: Kendall Whaling Museum Monograph Series No. 10, 1997.

Feyerabend, P. K. "On the Critique of Scientific Reasoning." In R. S. Cohen, P. K. Feyerabend, and M. W. Wartofsky, eds., *Essays in Memory of Imre Lakatos*, 109–43. Dordrecht: D. Reidel Publishing Company, 1976.

Feynman, R. P. *The Meaning of It All: Thoughts of a Citizen-Scientist*. Reading, MA: Perseus Books, 1998.

Flory, E. K. "Construing the Pelly and Packwood-Magnuson Amendments: The D.C. Circuit Court Harpoons Executive Discretion." *Washington Law Review*, vol. 61, no. 631 (1986): 631–36.

Freeman, M. M. R. "Science and Trans-Science in the Whaling Debate." In Freeman and Kreuter, *Elephants and Whales: Resources for Whom?* 143–57.

———. "Why Whale?" In O. D. Jonsson, ed., *Whales and Ethics*, 39–56. Reykjavik: Fisheries Research Institute, University of Iceland University Press, 1992.

Freeman, M. M. R., and S. R. Kellert, "International Attitudes to Whales, Whaling and the Use of Whale Products: A Six Country Survey." In Freeman and Kreuter, *Elephants and Whales: Resources for Whom?* 293–315.

Freeman, M. M. R., and U. P. Kreuter, eds. *Elephants and Whales: Resources for Whom?* Amsterdam: Gordon and Breach, 1994.

Friedheim, R. L. "Negotiating in the IWC Environment." In *Toward a Sustainable Whaling Regime*, 200–34. Seattle: University of Washington Press, 2001.

Gambell, R. "International Management of Whales and Whaling: An Historical Review of the Regulation of Commercial and Aboriginal Subsistence Whaling." *Arctic*, vol. 46, no. 2 (1993): 97–107.

———. "The International Whaling Commission Today." In G. Petursdottir, ed., *Whaling in the North Atlantic—Economic and Political Perspectives*. Reykjavik: University of Iceland, 1997. http//highnorth.no/Library/Publications/iceland/th-in-to.htm.

———. "The IWC and the Contemporary Whaling Debate." In J. R. Twiss Jr. and R. R. Reeves, eds., *Conservation and Management of Marine Mammals*, 179–97. Washington, DC: Smithsonian Institution Press, 1999.

Goodman, D. Letter in *Nature*, vol. 397 (January 28, 1999): 290.

———. "If Only Greenpeace Told the Truth about Whaling." *The Japan Times*, March 16, 2000.

Greenwood B. "Between Hope and a Hard Place." *Nature*, vol. 430 (August 2004): 926–27.

Gregoriadis, L. "Norwegian Fleet Sets Off to Kill 750 Minke in Biggest Slaughter of Whales for 14 Years." *The Independent View from Europe* (*Daily Yomiuri* supplement), May 9, 1999.

Grieco, J. M. "Anarchy and the Limits of Co-operation: A Realist Critique of the Newest Liberal Institutionalism." *International Organization*, vol. 42, no. 3 (1988): 485–508.

Gulland, J. "The End of Whaling?" *New Scientist*, October 29, 1988.

Haas, P. M. "Banning Chloroflurocarbons: Epistemic Community Efforts to Protect Stratospheric Ozone." In Haas, *Knowledge, Power, and International Policy Coordination*, 187–224.

———. "Do Regimes Matter? Epistemic Communities and Mediterranean Pollution Control." *International Organization*, vol. 43, no. 3 (1989): 377–403.

———. "Introduction: Epistemic Communities and International Policy Coordination." In Haas, *Knowledge, Power, and International Policy Coordination*, 1–35.

———, ed. *Knowledge, Power, and International Policy Coordination*. Columbia: University of South Carolina Press, 1997.

Haggard, S., and B. A. Simmons. "Theories of International Regimes." *International Organisation*, vol. 41, no. 3 (1987): 491–517.

Haverstock, H. [M. Heazle]. "Wails for Whales." *Japan International Journal*, vol. 3, no. 7 (July 1993): 34–39.

Heazle, M. "A Community on Edge." *Far Eastern Economic Review*, June 10, 1993.

———. "A Greener Shade of Whale?" *The Asian Wall Street Journal*, August 18, 1993.

———. "Whaling Town Taiji Dreams for Miracle IWC Reprieve." *The Japan Times*, May 23, 1999.

Heishman, S. J. "Behavioral-Cognitive Effects of Smoking." Paper presented at Addicted to Nicotine: A National Research Forum, Bethesda, MD, July 27–28, 1998. http://www.drugabuse.gov/meetsum/nicotine/heishman.html.

Hindess, B. *Philosophy and Methodology in the Social Sciences*. Sussex: The Harvester Press, 1977.

Holt, S. "A Comment on Tore Schweder's 'Protecting Whales by Distorting Uncertainty: Non-precautionary Mismanagement.'" *Fisheries Research*, vol. 52, no. 3 (July 2001): 227–30. Kalland, A. "Whose Whale is That?

Diverting the Commodity Path." In Freeman and Kreuter, *Elephants and Whales: Resources for Whom?* 159–86.

Keohane, R. O., and J. S. Nye. "Power and Interdependence Revisited." *International Organization*, vol. 41, no. 4 (1987): 725–53.

Klinowska, M. "Are Cetaceans Especially Smart?" *New Scientist*, October 29, 1988.

Kloppenberg, J. T. "Objectivity and Historicism: A Century of American Historical Writing." *American Historical Review*, vol. 94 (1989): 1011–30.

Knauss, J. A. "The International Whaling Commission—Its Past and Possible Future." *Ocean Development and International Law*, vol. 28 (1997): 79–87.

Kuhn, T. S. *The Structure of Scientific Revolutions*. Chicago: University of Chicago Press, 1996.

Larkin, P. A. "Objectives of Management." In R.T. Lackey and L.A. Nielsen, eds., *Fisheries Management*, 245–62. New York: John Wiley and Sons, 1980.

Larvor, B. *Lakatos: An Introduction*. New York: Routledge, 1998.

Lean, G., and R. Mendick. "Whale Population Devastated by Warming," *The Independent on Sunday*, July 29, 2001. Lemons, J. "The Conservation of Biodiversity: Scientific Uncertainty and the Burden of Proof." In Lemons, *Scientific Uncertainty and Environmental Problem Solving*, 206–31.

———, ed. *Scientific Uncertainty and Environmental Problem Solving*. Oxford: Blackwell Science, 1996.

Lloyd Parry, R. "Japan Admits Aid Deals Buy Support for Whaling." *The Independent*, July 19, 2001.

Lord, C. "Deep Feeling." *The Courier Mail*, April 25, 2000.

Ludwig, D., R. Hilborn, and C. Walters. "Uncertainty, Resource Exploitation, and Conservation: Lessons from History." *Science*, vol. 260 (April 2, 1993): 17–18.

MacDonald, J. M. "Appreciating the Precautionary Principle as an Ethical Evolution in Ocean Management." *Ocean Development and International Law*, vol. 26 (1995): 255–86.

Masood, E. "Error Re-opens 'Scientific' Whaling Debate." *Nature*, vol. 374 (April 13, 1995): 587.

McGuinness, P. P. "Sentimental Anti-whaling Lobby No Longer Cuts Much Ice." *The Australian*, July 1, 1992.

Melville, H. *Moby Dick*. New York: Bantam Books, 1981.

Murphy, K. "Deaths Signal Too Many Whales for the Ocean." *Los Angeles Times*, June 6, 1999.

Nicol, C. W. "Whaling Issue Being Used as a Political Tool." *Daily Yomiuri*, June 10, 1993.

Newsweek, "Song of the Whales," July 8, 1974.

Norris, C. *Against Relativism: Philosophy of Science, Deconstruction and Critical Theory*. Oxford: Blackwell Publishers, 1997.

Palmer, H. "U.K. Scientist Blames IWC for Ignoring Scientists' Unanimous Recommendation." *Daily Yomiuri*, June 10, 1993.

Pauly, D. *On the Sex of Fish and the Gender of Scientists: Collected Essays in Fisheries Science*. New York: Chapman and Hall, 1994.

Perry, S. L., D. P. DeMaster, and G. K. Silber. "Special Issue: The Great Whales: History and Status of Six Species Listed as Endangered Under the U.S. Endangered Species Act of 1973." *Marine Fisheries Review*, vol. 61, no. 1 (1999). http//spo.nwr.noaa.gov/mfr611/mfr611.htm.

Peterson, M. J. "Whalers, Cetologists, Environmentalists, and the International Management of Whaling." In Haas, *Knowledge, Power, and International Policy Coordination*, 147–86.

Popper, K. R. *The Logic of Scientific Discovery*. London: Hutchinson, 1972.

Preston, J., ed. *Paul K. Feyerabend: Knowledge, Science and Relativism*. Cambridge: Cambridge University Press, 1999.

Ravetz, J. R. *Scientific Knowledge and Its Social Problems*. New Brunswick, NJ: Transaction Publishers, 1996.

Ruppel Shell, E. "Resurgence of a Deadly Disease." *Atlantic Monthly*, August 1997.

Russell, B. *Sceptical Essays*. New York: Barnes and Noble, 1961.

Sanderson, I. T. *A History of Whaling*. New York: Barnes and Noble, 1993.

Sardar, Z. *Thomas Kuhn and the Science Wars*. Cambridge: Icon Books, 2000.

Sasamoto, H. "Japan Accused of Overhunting Small Whales." *Daily Yomiuri*, May 11, 1993.

Scarff, J. E. "The International Management of Whales, Dolphins, and Porpoises: An Interdisciplinary Assessment (Part One)." *Ecology Law Quarterly*, vol. 6, no. 323 (1977): 326–427.

Schweder, T. "Distortion of Uncertainty in Science: Antarctic Fin Whales in the 1950s." *Journal of International Wildlife Law and Policy*, vol. 3, no. 1 (2000): 73–92.

———. "Protecting Whales by Distorting Uncertainty: Non-precautionary Mismanagement?" *Fisheries Research*, vol. 52, no. 3 (July 2001): 217–25.

Shaw Bond, M., "Special Report: The Backlash Against NGOs," *Prospect*, April 2000.

Shrader-Frechette, K. "Methodological Rules for Four Classes of Scientific Uncertainty." In Lemons, *Scientific Uncertainty and Environmental Problem Solving*, 12–39.
Small, G. L. *The Blue Whale*. New York: Columbia University Press, 1971.
Smith, T. D. *Scaling Fisheries: The Science of Measuring the Effects of Fishing, 1855–1955*. Cambridge: Cambridge University Press, 1994.
Spencer, L. "The Not So Peaceful World Of Greenpeace." *Forbes*, November 11, 1991.
Stamps, T. Address at the closing session of the Third Pan-African Malaria Conference, Nairobi, Kenya, June 1998.
Stoett, P. *The International Politics of Whaling*. Vancouver: University of British Columbia Press, 1997.
Sweeney, K. "Ford Takes Brave Step on Global Warming." *Daily Yomiuri*, December 10, 1999.
Takahashi, J. "English Dominance in Whaling Debates: A Critical Analysis of Discourse in the International Whaling Commission." *Nichibunken Japan Review*, vol. 10 (1998).
Taverne, D. "Against Anti-science." *Prospect*, December 1999.
Thompson, P. B. "Uncertainty Arguments in Environmental Issues." *Environmental Ethics*, vol. 8 (Spring 1986): 59–75.
Timmins, N. "Japan Faces Trade War over Sperm Whale Hunt Curb." *The Times*, July 22, 1980.
———. "Fear of Sperm Whale's Extinction after Killing Ban Is Rejected." *The Times*, July 25, 1980.
———. "Catch Quotas Set for Most Whales." *The Times*, July 26, 1980.
Tønnessen, J. N., and A. O. Johnsen. *The History of Modern Whaling*. Berkeley: University of California Press, 1982.
Trigg, G. "Japanese Scientific Whaling: An Abuse of Right or Optimum Utilisation?" *Asia Pacific Journal of Environmental Law*, vol. 5, no. 1 (2000): 33–59.
Whymant, R. "Japan Launches Secretive Whale Hunt 'For Science.'" *The Times*, July 6, 1996.
Wilkinson, P. "The World Needs More Than Protests." *Nature*, vol. 396 (December 10, 1998): 511–12.
Young, O. R. *International Cooperation: Building Regimes for Natural Resources and the Environment*. Ithaca: Cornell University Press, 1989.
———. *International Governance: Protecting the Environment in a Stateless Society*. Ithaca: Cornell University Press, 1994.
Young, O. R., G. J. Demko, and K. Ramakrishna, eds. *Global Environmen-*

tal Change and International Governance. Hanover: University Press of New England, 1996.

Young, O. R., and G. Osherenko. "The Formation of International Regimes: Hypotheses and Cases." In O. R. Young and G. Osherenko, eds., *Polar Politics*, 1–21.

———, eds. *Polar Politics: Creating International Environmental Regimes*. Ithaca: Cornell University Press, 1993.

IWC REPORTS AND DOCUMENTS

Each of the IWC report entries shows two different years. The first, in parentheses, is the year of the associated annual meeting, as described by the Chairman's Report. The second, following the place of publication, is the year the report was published; in both the notes and the bibliography, an IWC document's year of publication is never enclosed in parentheses. The year of publication for each report is the year after the annual meeting, except for the annual meetings held from 1962 to 1974, when publication of each report occurred two years after the associated annual meeting. The commission's 1975 and 1976 annual meetings are contained in one report, the *Twenty-seventh Report* published in 1977, which means that the annual reports since that year were again published the year subsequent to the associated annual meetings.

All of the IWC references in the notes and bibliography include the report number and year of publication except for those documents that either are not a complete annual report or do not reference one in the note itself. Examples of the latter include opening statements, verbatim reports, and some individual committee reports and papers that were not published with the relevant annual report. In these instances, notes and bibliography entries indicate the meeting itself in one of two ways: as either IWC third meeting (1951); or SC/44/ . . . , where 44 indicates the forty-fourth meeting.

IWC ANNUAL REPORTS

IWC. *First Report of the Commission* (1949). London, 1950.
———. *Second Report of the Commission* (1950). London, 1951.
———. *Third Report of the Commission* (1951). London, 1952.
———. *Fourth Report of the Commission* (1952). London, 1953.
———. *Fifth Report of the Commission* (1953). London, 1954.

———. *Sixth Report of the Commission* (1954). London, 1955.
———. *Seventh Report of the Commission* (1955). London, 1956.
———. *Eighth Report of the Commission* (1956). London, 1957.
———. *Ninth Report of the Commission* (1957). London, 1958.
———. *Tenth Report of the Commission* (1958). London, 1959.
———. *Eleventh Report of the Commission* (1959). London, 1960.
———. *Twelfth Report of the Commission* (1960). London, 1961.
———. *Thirteenth Report of the Commission* (1961). London, 1962.
———. *Fourteenth Report of the Commission* (1962). London, 1964.
———. *Fifteenth Report of the Commission* (1963). London, 1965.
———. *Sixteenth Report of the Commission* (1964). London, 1966.
———. *Seventeenth Report of the Commission* (1965). London, 1967.
———. *Eighteenth Report of the Commission* (1966). London, 1968.
———. *Nineteenth Report of the Commission* (1967). London, 1969.
———. *Twentieth Report of the Commission* (1968). London, 1970.
———. *Twenty-first Report of the Commission* (1969). London, 1971.
———. *Twenty-second Report of the Commission* (1970). London, 1972.
———. *Twenty-third Report of the Commission* (1971). London, 1973.
———. *Twenty-fourth Report of the Commission* (1972). London, 1974.
———. *Twenty-fifth Report of the Commission* (1973). London, 1975.
———. *Twenty-sixth Report of the Commission* (1974). London, 1976.
———. *Twenty-seventh Report of the International Whaling Commission* (1975 and 1976). Cambridge, 1977.
———. *Twenty-eighth Report of the International Whaling Commission* (1977). Cambridge, 1978.
———. *Twenty-ninth Report of the International Whaling Commission* (1978). Cambridge, 1979.
———. *Thirtieth Report of the International Whaling Commission* (1979). Cambridge, 1980.
———. *Thirty-first Report of the International Whaling Commission* (1980). Cambridge, 1981.
———. *Thirty-second Report of the International Whaling Commission* (1981). Cambridge, 1982.
———. *Thirty-third Report of the International Whaling Commission* (1982). Cambridge, 1983.
———. *Thirty-fourth Report of the International Whaling Commission* (1983). Cambridge, 1984.
———. *Thirty-fifth Report of the International Whaling Commission* (1984). Cambridge, 1985.

———. *Thirty-sixth Report of the International Whaling Commission* (1985). Cambridge, 1986.
———. *Thirty-seventh Report of the International Whaling Commission* (1986). Cambridge, 1987.
———. *Thirty-eighth Report of the International Whaling Commission* (1987). Cambridge, 1988.
———. *Thirty-ninth Report of the International Whaling Commission* (1988). Cambridge, 1989.
———. *Fortieth Report of the International Whaling Commission* (1989). Cambridge, 1990.
———. *Forty-first Report of the International Whaling Commission* (1990). Cambridge, 1991.
———. *Forty-second Report of the International Whaling Commission* (1991). Cambridge, 1992.
———. *Forty-third Report of the International Whaling Commission* (1992). Cambridge, 1993.
———. *Forty-fourth Report of the International Whaling Commission* (1993). Cambridge, 1994.
———. *Forty-fifth Report of the International Whaling Commission* (1994). Cambridge, 1995.
———. *Forty-sixth Report of the International Whaling Commission* (1995). Cambridge, 1996.
———. *Forty-seventh Report of the International Whaling Commission* (1996). Cambridge, 1997.
———. *Forty-eighth Report of the International Whaling Commission* (1997). Cambridge, 1998.
———. *Annual Report of the International Whaling Commission* (1998). Cambridge, 1999.
———. *Annual Report of the International Whaling Commission* (1999). Cambridge, 2000.
———. *Annual Report of the International Whaling Commission* (2000). Cambridge, 2001.

ADDITIONAL IWC PUBLICATIONS

"Comprehensive Assessment Workshop on Catch Per Unit Effort (CPUE)." In Donovan, *Comprehensive Assessment of Whale Stocks*, 15–19.

"Comprehensive Assessment Workshop on Management." In Donovan, *Comprehensive Assessment of Whale Stocks*, 21–28.

"The Condition of the Antarctic Whale Stocks." Appendix to the "Report of the Special Scientific Sub-Committee." IWC fifth meeting (1953), Document II.

Cooke, J. "Simulation Studies of Two Whale Stock Management Procedures." In Donovan, *Comprehensive Assessment of Whale Stocks*, 147–48.

Donovan, G. P., ed. *The Comprehensive Assessment of Whale Stocks: The Early Years*, Special Issue 11. Cambridge: IWC, 1989.

———. Preface. In Donovan, *Comprehensive Assessment of Whale Stocks*.

"Draft Report of the Scientific Committee." IWC sixth meeting (1954), Document XV.

Hammond, P. "The Conservation and Management of Small Cetaceans: A Review of the Options." IWC Scientific Committee Paper SC/43/SM12.

———. Letter of resignation to the IWC, May 26, 1993, IWC Ref: IWC/2.1.

Mackintosh, N. A. "Part 1: Research on the Natural History of Whales." IWC third meeting (1951), Document III.

———. "Part 2: Memorandum on the Southern Stocks of Humpback Whales." IWC third meeting (1951), Document III.

———. "General Scientific Report." IWC sixth meeting (1954), Document III.

"Memorandum of the Dutch Delegation on the Reduction of the 16,000 Blue Whale Units to the IWC Scientific Committee." IWC fifth meeting (1953).

"Opening Statement by the Australian Commissioner." IWC thirty-first meeting (1979), London.

"Opening Statement by New Zealand." IWC forty-sixth meeting (1994), Puerto Vallarta.

"Opening Statement of New Zealand Commissioner B. J. Lynch." IWC thirty-first meeting (1979), London.

"Opening Statement of Richard A. Frank, U.S. Commissioner." IWC thirty-first meeting (1979), London.

"Report of the Scientific Committee." IWC fourth meeting (1952), Document XV.

"Report of the Scientific Committee." IWC fifth meeting, (1953), Document XVII.

"Report of the Special Scientific Sub-Committee." IWC sixth meeting (1954), Document II.

"Report of the Scientific Sub-Committee." IWC seventh meeting (1955), Document III.

"Report of the Special Meeting of the Scientific Committee on Planning for a Comprehensive Assessment of Whale Stocks." In Donovan, *The Comprehensive Assessment of Whale Stocks*, 3–13.

"Report of the Technical Committee." IWC seventh meeting (1955), Document XIX.

"Revised Rules of Procedure of the Scientific Committee." Annex M, Report of the Scientific Committee. *Journal of Cetacean Research and Management*, vol. 1 (1999): 247–50.

"Revised Rules of Procedure to Incorporate Changes Agreed at IWC/53." Circular Communication to Commissioners and Contracting Governments IWC.CCG.209. IWC document NJG/JAC/29093, October 9, 2001.

Schweder, T. *Intransigence, Incompetence or Political Expediency? Dutch Scientists in the International Whaling Commission in the 1950s: Injection of Uncertainty*. IWC Scientific Committee Paper SC/44/0 13, 1992.

"Verbatim Report of the Plenary Sessions." IWC third meeting (1951), Document XII.

"Verbatim Report of the Plenary Sessions." IWC sixth meeting (1954), Document XIV (D).

"Verbatim Report of the Plenary Sessions." IWC seventh meeting (1955), Document XXL-C.

"Verbatim Report of the Plenary Sessions." IWC eighth meeting (1956), Document XIIIC.

"Verbatim Report of the Plenary Sessions." IWC tenth meeting (1958), Document XIII.

"Verbatim Report of the Plenary Sessions." IWC twelfth meeting (1960), IWC/12/11.

"Verbatim Report of the Plenary Sessions." IWC fifteenth meeting (1963), IWC/15/17.

"Verbatim Report of the Plenary Sessions." IWC sixteenth meeting (1964), IWC/16/15.

"Verbatim Report of the Plenary Sessions." IWC seventeenth meeting (1965), IWC/17/13.

GOVERNMENT, GOVERNMENT ORGANIZATION, AND NGO DOCUMENTS AND PUBLICATIONS

Anderson, R. M., R. J. H. Beverton, A. Semb-Johansson, and L. Walløe. *The State of the North-East Atlantic Minke Whale Stock: Report of the Group* of Scientists Appointed by the Norwegian Government to Review the Basis for Norway's Harvesting. Oslo: Okoforsk, 1987.

Australian Department of the Environment and Heritage. "A Universal Metaphor: Australia's Opposition to Commercial Whaling." Report of the National Task Force on Whaling. May 1997. http://www.deh.gov.au/coasts/species/cetaceans/whaling/.

Blichfeldt, G., ed. *11 Essays on Whales and Man*. Lofoten, Norway: High North Alliance, 1994.

Butterworth, D. S. "Sustainable Utilisation of Marine Mammal Resources." In Hallenstvedt and Blichfeldt, *Additional Essays on Whales and Man*, 11–13.

Donovan, G. P. "The International Whaling Commission and the Revised Management Procedure." In Hallenstvedt and Blichfeldt, *Additional Essays on Whales and Man*, 4–10.

Elk, C., and E. H. Buck. *Norwegian Commercial Whaling: Issues for Congress*. Washington, DC: Congressional Research Service, December 31, 1996.

Extracts and Summary from the Danish Environmental Protection Agency's Conference on the Precautionary Principle. Copenhagen: Ministry of Environment and Protection, Danish Environmental Protection Agency, May 29, 1998.

Hallenstvedt, E., and G. Blichfeldt, eds. *Additional Essays on Whales and Man*. Lofoten, Norway: High North Alliance, 1995.

Harremoes, P. "Can Risk Analysis Be Applied in Connection with the Precautionary Principle?" In *Extracts and Summary from the Danish Environmental Protection Agency's Conference on the Precautionary Principle*, 30–35.

Holt, S. "The Whaling Game." *The Siren*, no. 35 (December 1987): 3–6. (Published by UNEP's Oceans and Coastal Areas Programme.)

———. *Minkes' Saga, Part II or Whatever Happened to Vesterfjord?* IFAW/IOI Technical Briefing, vol. 93, no. 7 (May 1993).

———. "Coverage of the International Whaling Commission's 1996 Annual Meeting June 23–28, Aberdeen, Scotland." *Eco*, no. 3 (June 26, 1996).

Institute of Cetacean Research. *Whale Management Under the International Whaling Commission's Revised Management Procedure (RMP)*. http://www.whalesci.org/research/management.html.

———. *Papers on Small-Type Coastal Whaling Submitted by the Government of Japan to the International Whaling Commission 1986–1995*. Tokyo: Institute of Cetacean Research, 1996.

———. *Research on Whales*. Tokyo: Institute of Cetacean Research, 1997.

The International Harpoon, no. 2, June 25, 1996. (Published by the High North Alliance.)

Kalland, A. "Super Whale: The Use of Myths and Symbols in Environmentalism." In Blichfeldt, *11 Essays on Whales and Man*, 39–41.

Kasamatsu, F. "Counting Whales in the Antarctic." In Institute of Cetacean Research, *Research on Whales*, 29–38.

Klinowska, M. "Brains, Behaviour and Intelligence in Cetaceans (Whales, Dolphins and Porpoises)." In Blichfeldt, *11 Essays on Whales and Man*, 21–26.

Parsons, R. M. *The US Marine Mammal Protection Act: A Report to the High North Alliance on the Waiver Process and Exemptions to the Embargo on Marine Mammal Products*. http://www.highnorth.no/Library/Trade/GATT_WTO/us-ma-ma.htm. Report contracted by the High North Alliance in February 1996; released in November 1997.

Sorensen, H. "The Environment Movement and Minke Whaling." In Blichfeldt, *11 Essays on Whales and Man*, 27–30.

PERSONAL CORRESPONDENCE AND INTERVIEWS

Althaus, Dr. Thomas. IWC commissioner for Switzerland. Interviewed in London, July 2001.

Aron, Dr. William. Former U.S. IWC commissioner and Scientific Committee member. Personal correspondence.

———. Interviewed in Tokyo, March 2001.

Asmundsson, Stefan. IWC commissioner for Iceland. Interviewed in London, July 2001, and Shimonoseki, Japan, May 2002.

Bannister, John. IWC Scientific Committee member. Personal correspondence.

———. Interviewed in London, July 2001.

Berninger, Matthias. IWC commissioner for Germany. Interviewed in London, July 2001.

Best, Dr. Peter. Former IWC Scientific Committee member. Personal correspondence.

Butterworth, Professor Doug. IWC Scientific Committee member. Personal correspondence.

———. Interviewed in Tokyo, March 2001, and London, July 2001.

Canny, Michael. Republic of Ireland IWC commissioner. Interviewed in London, July 2001.

Cooke, Dr. Justin. IWC Scientific Committee member and World Conservation Union representative. Interviewed in London, July 2001, and Shimonoseki, Japan, May 2002.

Cowan, Richard. UK IWC commissioner. Interviewed in London, July 2001.

DeMaster, Dr. Doug. IWC Scientific Committee chairman-elect (2002). Personal correspondence.

———. Interviewed in Shimonoseki, Japan, May 2002.

Donovan, Greg. IWC scientific editor. Personal correspondence.

———. Interviewed in London, July 2001, and Shimonoseki, Japan, May 2002.

Forkan, Patricia. The Humane Society of the United States. Interviewed in London, July 2001.

Freeman, Professor Milton M. R. University of Alberta. Personal correspondence.

Gambell, Dr. Ray. Former IWC general secretary and Scientific Committee member. Personal correspondence.

———. Interviewed in London, July 2001.

Goodman, Dan. Advisor to the Institute of Cetacean Research, Tokyo, and former Canadian government delegate to the IWC. Personal correspondence.

———. Interviewed in Tokyo, March 2001.

Grande, Dr. Nicola. IWC Secretary. Interviewed in Histon, UK, July 2001.

Harvey, Martin. IWC executive officer. Interviewed in Shimonoseki, Japan, May 2002.

Hill, Senator Robert. Australian Minister for the Environment and Heritage. Interviewed in London, July 2001.

Holt, Dr. Sidney. Former FAO scientist and member of the Committee of Three and Committee of Four. Interviewed in London, July 2001.

Hutton, Dr. Jon. University of Cambridge and Africa Resources Trust. Interviewed in Histon, UK, July 2001.

Komatsu, Masayuki. Alternative IWC commissioner for Japan. Interviewed in London, July 2001.

Lambertsen, Dr. Richard. Former IWC Scientific Committee member. Personal correspondence.

Lapointe, Eugene. World Conservation Trust and former secretary-general for the Convention on International Trade in Endangered Species. Interviewed in London, July 2001.

Mclay, Jim. Alternative IWC commissioner for New Zealand. Interviewed in London, July 2001.

Morishita, Joji. Deputy-Director, Far Seas Fisheries Division—Government of Japan. Interviewed in London, July 2001.

Ohsumi, Dr. Seiji. Director-General, Institute of Cetacean Research, Tokyo, and IWC Scientific Committee member. Interviewed in Tokyo, March 2001.

Paul, Jonathan. Ocean Defence International. Interviewed in London, July 2001.

Phillips, Cassandra. World Wildlife Fund. Interviewed in London, July 2001.

Pombo, Richard. U.S. congressman, 11th district, California. Interviewed in Shimonoseki, Japan, May 2002.

Schweder, Professor Tore. IWC Scientific Committee member. Personal correspondence.

Spong, Dr. Paul. Earth Island Institute. Personal correspondence.

Van Note, Craig. Monitor. Interviewed in London, July 2001.

Walløe, Professor Lars. IWC Scientific Committee member. Interviewed in London, July 2001.

White, Ben. Animal Welfare Institute. Interviewed in London, July 2001.

INDEX

aboriginal whaling, 203*n*1, 229*n*53
Ad Hoc Scientific Committee (AHSC), 80–86, 88–89, 90–91, 92–93
age determination methods, 82, 92–93, 218*n*76
Age Determination Sub-committee, 82–83
Allen, Kay Radway, 55, 81, 84, 130, 145, 152, 204*nn*3-4
Andresen, Steinar, 47, 153
Antarctic sanctuary area: opening of, 58–59; pre-IWC agreement, 209*n*10
Aron, William, 145, 148, 178, 225–26*n*18, 226*n*29, 229*n*51
Australia: and Committee of Three, 80; humpback whales, 99, 213*n*55; hunting cessation, 229*n*53; moratorium, 158, 169, 232*n*94; NMP proposal, 145; protectionist approach, 174; public opinion poll, 233*n*101; quota votes, 99, 119

Baleana, 78
baleen whales, families of, 204*n*3. *See also specific whale families, e.g.,* fin whales; sei whales
bans, hunting: Australia, 229*n*53; blue whale debate, 100–106, 121, 127–28; humpback whales, 54, 56–57, 99, 213*n*55; Scientific Committee recommendation, 127–28
bargaining, regime theory, 211*n*34
Beard, Charles, 207*n*35

Beddington, John, 150, 229*n*51
benefit-risk factor. *See* utility factor, generally
Bentham, Jeremy, 22–23
Bergersen, Birger, 53, 55–56
Best, Peter, 143
Beverton, Raymond, 49
Birnie, P., 64, 104, 215*n*13
blue whales: catch statistics summarized, 192–97; in Mysticeti group, 204*n*3; during the 1950s, 55–57; and technological innovation, 38–39
blue whales, during the 1960s: hunting ban debate, 100–106, 121, 127–28; pygmy blue population, 100–106, 112, 120–23; Scientific Committee recommendations, 103, 112; scientific warnings, 88–89, 95–97; season change debate, 90
Blue Whale Unit (BWU): adoption, 40, 53–54, 208*n*4, 209*n*13; at London Conference, 210*n*19; during the 1970s, 137; statistics summarized, 189–90
Blue Whale Unit (BWU), during the 1950s: IWC votes, 58–59, 61–62; in national quotas discussion, 70–71, 73–77, 217–18*n*55; objections/vetoes, 59, 60–61; Scientific Committee recommendations, 50, 54–55
Blue Whale Unit (BWU), during the 1960s: FAO recommendations, 127; IWC votes, 78, 99, 121, 128; in national quotas discussion, 77–78,

253

254   INDEX

Blue Whale Unit (BWU) *(continued)* 86–88, 222*n*39; reduction debates, 98–99, 112–13, 116–20; science-based pledge vote, 111, 220*n*4; scientific committees' recommendations, 84–85, 96–97, 98, 115–16, 117, 125, 129–30; transfers of, 97–98, 104–5
Bodansky, Daniel, 157
bowhead whales, in Mysticeti group, 204*n*3
Brazil, 225*n*15
Bryde's whales, 137, 139–40, 142, 192–97, 204*n*3
Butterworth, Doug, 162–63, 174, 176, 182–83
BWU. *See* Blue Whale Unit (BWU) entries

Canada, 59, 99, 125, 152
cancer cure example, 34–35, 177
Carr, E. H., 13
catching effort measurement, 82–83, 85
catch limit algorithm, Revised Management Procedure, 159, 160–65
catch limits. *See* quotas *entries*
catch statistics: during the *1950s*, 56; during the *1960s*, 88, 97, 99, 101, 102, 113, 124, 128–29, 222*n*45, 223*n*56; during the *1970s*, 143, 226*n*22; Soviet Union falsification, 88, 138, 220*n*7, 224–25*n*13; summarized, 189–90, 192–97; World War II, 209–10*n*16
Chalmers, Alan, 9
Chapman, Douglas, 81, 84, 130, 142–43, 151, 152, 227–28*n*41
Chile, 225*n*15
Clark, Colin W., 212–13*n*49
cognitive theory, 181–83
Committee of Four (COF): commission reception, 111–12, 117; population estimates, 108–9, 113–17, 222*n*40; quota recommendations, 103, 112–13
Committee of Six, 124–25
Committee of Three (COT): commission reception, 111–12; data organization challenge, 83; formation/responsibilities, 79, 80, 81–83; funding, 83–84, 90, 217*n*49; mortality rate calculations, 92–93; MSY calculations/recommendations, 93–97; population

warnings, 91–92; pygmy blue whale, 101–2; quota recommendations, 84–85, 99
comprehensive assessment, Revised Management Procedure, 156–57, 159–60
conservationists, 170
Cooke, Justin, 150–51, 162, 164
CPUE indicators, 55–56, 84–86, 93, 108–9
criterion I, defined, 21. *See also* utility factor, generally
criterion II, defined, 27–28. *See also* utility factor, generally
Cushing, David H., 51, 64

de la Mare, Bill, 150
De Lury method, 93
Denmark, 99, 145
Dobson, A. T. A., 61
Doi, T., 101, 102, 130
Donovan, Greg, 146, 158, 161, 167, 230*n*66
Doyal, Len, 24–25
Drion, E. F., 58, 62, 73, 216*n*19

ear plug measurement, 82, 92–93
Ebola virus example, 28–29
economic factors: before/during World War II, 37–38; in BWU reduction debate, 98–99, 100; decline of, 134–35, 155, 223*n*1; and national quotas discussions, 63, 67–68, 69, 74, 75–76; postwar investments, 39–41, 52
Elliot, Gerald, 54, 64, 72–73, 104, 215*n*13, 224–25*n*13
empiricism epistemology, 11–21, 31–32
Endangered Species Act, 223–27*n*6
environmental groups, 143–44, 150–51, 154, 169, 170–75
Environmental Protection Agency, 225–26*n*18
epistemic communities, 181, 182, 234*n*3
escape clause, in ICRW, 45–47. *See also* objections

factory ships: Dutch increase, 60; as economic pressure, 40–41; in national quotas negotiations, 71, 73; observer proposal, 91, 138; sales of, 78, 97–

98, 104–6, 118; Soviet Union increase, 71, 78; statistics summarized, 189–90
false positive *vs.* false negative errors, acceptance, 166
falsification theory, 11–12, 14–21
FAO (Food and Agriculture Organization), 125–29, 152, 222*n*40, 226–27*n*32
Fauna Preservation Society, 144
Feyerabend, Paul, 11, 12
fin whales: catch statistics summarized, 192–97; in Mysticeti group, 204*n*3; during the *1950s*, 55–58, 92; and technological innovation, 38–39
fin whales, during the *1960s*: catch statistics, 97, 114, 128–29, 223*n*56; FAO recommendations, 127–29; population estimates, 116, 130; science warnings, 95–97; scientific committees' recommendations, 103, 112, 113–14, 128; scientific warnings, 80–81, 88–89
fin whales, during the *1970s*: catch statistics, 226*n*22; NMP adoption, 146–47; population estimates, 130–31; quota debates, 141–42; species-based quota adoption, 137
Food and Agriculture Organization (FAO), 125–29, 152, 222*n*40, 226–27*n*32
France, 61, 99, 125, 174
Frank, Richard A., 151–52, 168, 229*n*53
Frost, Sidney, 229*n*53
Fujita, I., 117, 123
funding: IWC generally, 225*n*15; scientific committees, 51, 81, 83–84, 90, 217*n*49
fundraising strategies, environmental groups, 172–73

Gambell, Ray, 47, 76–77, 147, 150–51, 212*n*39, 220*n*7
Geneva Convention for the Regulation of Whaling, 208*n*4, 209*n*10
Germany, 41, 208*n*4, 233*n*101
GMF controversy, 29–30
Gough, Ian, 24–25
Greenpeace, 170, 172
grey whales, 38, 204*n*3, 209*n*10
Gulland, John, 91, 103, 152, 173, 230*n*61

Haas, P. M., 181, 182, 234*nn*3–4
Hammond, Philip, 110, 234*n*5
harpoon guns, 38–39
heuristic approach, scientific discovery, 17–18
Hindes, Barry, 15–16
history, utility factor, 26–27, 207*n*35
Hjort, Johan, 93
Holt, Sidney: as COT member, 81, 84; on FAO conditions, 125–26; moratorium support, 150–51, 152; NMP criticisms, 149–51; pygmy blue whale debate, 123; RMP calculations, 163; statistical methodology development, 49; on uncertainty argument, 109, 229*n*51
Hume, David, 13
humpback whales: catch declines, 55; catch statistics summarized, 192–97; hunting ban, 54, 56–57, 99, 213*n*55; in Mysticeti group, 204*n*3; in *1950s* quota debates, 54; pre-IWC agreements, 209*n*10; scientific warnings, 88–89, 95–97; season changes, 90
hydrogenation technology, introduction, 208–9*n*9

Iceland, 152, 228*n*46, 231*n*73
Ichihara, T., 102
ICRW (International Convention for the Regulation of Whaling), 42, 44–48, 63, 70, 158
Indian Ocean Sanctuary, 151, 154
inductive reasoning, 13–15
International Convention for the Regulation of Whaling (ICRW), 42, 44–48, 63, 70, 158
International Decade of Cetacean Research, 138–39, 147
International Observer Scheme, 91, 125, 138, 159–60, 221*n*31
International Society for the Protection of Animals, 143–44
International Whaling Commission (IWC), overview, 3–7, 42, 44–48. *See also specific topics, e.g.,* quotas *entries;* Scientific Committee *entries*

Japan: catch statistics, 97, 98, 101, 114, 124, 128, 143, 222*n*45; and Commit-

256 INDEX

Japan *(continued)*
  tee of Three, 80, 84; economic factors, 69, 98–99, 226*n*29; moratorium proposal, 152–53, 228*n*46; postwar industry development, 41, 43, 211*n*25; pre-WWII agreements, 208*n*4; public opinion poll, 233*n*101; RMP agreement, 231*n*73; scientific research premise, 174; ship purchases, 97–98, 104–6, 124; and superwhale concept, 175
Japan, 1950s quota debates: IWC withdrawal, 62, 71, 75; national-based, 63–64, 72; votes/objections, 58, 59, 61, 62, 75
Japan, 1960s quota debates: argument themes, 98, 112, 116, 117, 122–24, 131–32; national-based, 86–88, 222*n*39; observer demand, 125; pygmy blue whales, 100–106, 120–23; science-based pledge refusal, 111, 220*n*4; suspension proposal, 78; votes, 78, 119, 121, 220*n*4
Japan, 1970s quota debates: agreements, 131; argument themes, 142; minke whales, 141–43, 226*n*29; votes/objections, 141–42, 144, 224*n*9
Johnsen, Arne Odd, 47, 53, 64, 71, 74, 105–6, 209–10*n*16, 215*n*13, 216*n*27

Kalland, Arne, 171–72, 174
Kellogg, Remington, 53, 101
Kloppenberg, James T., 207*n*35
Kuhn, Thomas, 11–12, 16–21

Lakatos, Imre, 11–12, 16, 17–21
Lamberson, Roland, 212–13*n*49
land station catches, 99, 213*n*55, 221*n*31
Lankaster, Kees, 150
Law, R. M., 57–58, 62, 80, 92
Lemons, John, 166, 167
Lienesch, G. J., 60
The Logic of Scientific Discovery (Popper), 14
London meetings: during the 1940s, 42, 210*n*19; during the 1950s, 55, 60–61, 63, 70, 74–75; during the 1960s, 77, 80–81, 85–87; during the 1970s, 136, 141, 151

Mackintosh, N. A., 50, 53, 54, 56
malaria example, 28–29
maps, 199–201
Marine Mammal Protection Act, 223–27*n*6
marketing value, whales, 171–74
maximum sustainable yield calculations: in NMP, 227–28*n*41; in RMP, 159–66; during the 1960s, 93–97, 101–2, 113, 116; during the 1970s, 139–40, 224*n*9
measurement subcommittee, 82–83
medicine examples, 21–22, 28–29, 34–35
minke whales: in cancer cure example, 34–35, 177; catch statistics, 143, 192–97, 222*n*45; Greenpeace obituary, 172; in Mysticeti group, 204*n*3; NMP adoption, 146–47; population estimates, 142–43, 165; quota debates, 141–42, 226*n*29; votes/objections, 142
moratoria proposals: aboriginal whaling distinction, 203*n*1; adoptions, 151, 152, 156, 230*n*66; as ethical concern, 169, 232*n*94; IWC votes, 141; member opposition arguments, 152–53, 228*n*46, 230*n*61; and NMP-based management, 146, 147–48; as research opportunity, 156–57; as science-based decision, 133–35; Scientific Committee responses, 138–39, 141, 149, 150–51, 156–57, 227*n*37, 228*nn*46–47; U.S. efforts, 136–37, 139, 140, 141, 151–52, 153, 227*n*37
mortality rate calculations: during the 1950s, 57–58; during the 1960s, 92–93
Moscow meetings, 58, 60, 73, 91
MSY calculations. *See* maximum sustainable yield calculations

National Environmental Policy Act, 225–26*n*18
National Oceanic and Atmospheric Administration, 225–26*n*18
national quotas: overview, 68–70, 74, 217–18*n*55; allocation discussions, 63–64, 70–75, 222*n*39; failure factors, 75–76; final agreements, 86–88; incentives for, 63, 66–67; and member withdrawals, 62–63, 71, 74–75, 77–78; as prewar policy, 62–63

needs, in utility factor, 27–30
needs *vs.* wants, judgment dilemma, 23–27
negotiations, regime theory, 211*n*34
Netherlands: catch statistics, 88, 97, 98; economic factors, 67, 69, 76, 98; industry development, 42–43; ship sales, 105–6
Netherlands, *1950s* quota debates: CPUE indicators, 55, 56; IWC withdrawal, 62, 71, 76; national-based, 63–64, 70–75, 76–77, 215*n*13; votes/objections, 58, 59, 60–62, 214*n*79
Netherlands, *1960s* quota debates: argument themes, 98–99; IWC withdrawal, 77–78, 87–88; national-based, 86–88; science-based pledge refusal, 111, 220*n*4; ship sales impact, 118; votes/objections, 90, 220*n*4
New Management Procedure (NMP): adoption, 145–47, 227–28*n*41; implementation challenges, 147–51; and moratorium momentum, 152, 168–69; RMP compared, 164–65
New Zealand: and Committee of Three, 80; moratorium, 158, 169, 187, 232*n*94; protectionist approach, 174; in *1950s* quota debates, 50; in *1960s* quota debates, 89, 99
NGOs, 143–44, 150–51, 154, 169, 170–75
Nixon administration, 225–26*n*18
NMP. *See* New Management Procedure (NMP)
Norway: catch statistics, 97, 124, 128, 129; economic factors, 43–44, 67–68, 98; moratorium proposal, 152–53, 228*n*46; observer proposal, 91, 221*n*31; pre-IWC meetings/agreements, 41–42, 208*n*4, 210*n*18; public opinion poll, 233*n*101; resistance to postwar newcomers, 41, 42–43, 210*n*21; RMP agreement, 231*n*73; sales of ships, 97, 124; whaling technologies, 38–39; World War II whaling, 41, 209–10*n*16
Norway, *1950s* quota debates: IWC withdrawal, 62–63, 71–72, 74–75, 76; national-based, 63–64, 74–75; votes/objections, 59, 61, 62, 75
Norway, *1960s* quota debates: compromise proposals, 118, 125; IWC withdrawal, 77–78, 87–88; national-based, 86–88, 217–18*n*55, 222*n*39; votes, 99, 119
Norwegian Crew Law, 42–43

objections: during the *1950s*, 59, 60–62, 73, 214*n*79; during the *1960s*, 78, 90; during the *1970s*, 141–42, 144. *See also* votes
observer proposal, 91, 125, 138, 159–60, 221*n*31
Ohsumi, Seiji, 142–43
Omura, H., 103, 122–23
Onassis, Aristotle, 210*n*21
Osherenko, Gail, 211*n*34
Oslo meeting, 74
Ottestad, P., 58, 218*n*76

Packwood-Magnuson Amendment, 223–27*n*6
Panama, 58, 59
Pelly Amendment, 223–27*n*6
Peru, 152–53, 225*n*15
Peterson, M. J., 183
pharmaceutical examples, 28–29
philosophy of science, historical developments, 11–21
political costs, whaling support, 154–55, 166
Popper, Karl, 11–12, 14–21
population estimates: with RMP, 159–66; during the *1950s*, 55–56, 57–58; during the *1960s*, 101, 113, 114–16, 129–30; *vs.* limitation principle, 53–54. *See also* maximum sustainable yield calculations
Portugal, 225*n*15
precautionary principle: emergence in whaling debate, 117–18, 123–24; in GMF controversy, 29–30; implementation difficulties, 30–31, 167–68; and moratorium adoption, 153–54; and RMP development, 154–59, 166–68, 174–75; and scientific method, 7–10; and utility factor generally, 29–33
pre-IWC whaling, 37–42, 208*n*4, *n*9, 209*n*10, *n*13, *n*16, 210*nn*18–19
preservationists, 170–75

258 INDEX

prices, whale products, 39–40, 191, 222*n*45
public opinion, 169, 233*n*101
pygmy blue whales, 100–106, 112, 120–23

quotas: in Blue Whale Unit adoption, 209*n*13; maps for, 199–201
quotas, during the *1940s*: humpback whales, 213*n*55; and limitation principle, 53–54; as whaling olympic impetus, 40
quotas, during the *1950s*: IWC votes, 56, 58–59, 61, 73, 75; objections/vetoes, 59, 60–62, 73, 214*n*79; Scientific Committee recommendations, 50, 54–56, 57, 58, 66–67. *See also* national quotas
quotas, during the *1960s*: after member withdrawals, 77–78; decision deferrals, 89–90, 111–13; FAO recommendations, 125–29; IWC votes, 78, 99, 119–20, 128, 220*n*4; and precautionary principle emergence, 117, 123–24; pygmy blue whale, 100–106; reduction debates, 98–99, 125; Scientific Committee recommendations, 84–85, 96, 98, 125; suspension proposal, 77–79; transfers of, 78, 97–98, 104–5, 124; voluntary limit agreement, 120
quotas, during the *1970s*: agreements, 131, 137; NGO emergence, 143–44; NMP adoption, 146–47, 227–28*n*41; votes/objections, 141–42, 144, 224*n*9

rationalism epistemology, 11–21
regime theory, 181–83, 211*n*34, 234*n*3
Revised Management Procedure (RMP): development of, 154–59, 166–68, 174–75; elements, 159–66, 231*n*73
right whales, 38, 39, 204*n*3
risk factor. *See* utility factor, generally
Russell, Bertrand, 13–14, 19–20
Ruud, J., 56, 57–58, 92

Sandefjord meeting, 111–13
Sardar, Ziauddin, 19
Scarff, James E., 144, 215*n*13
Schaefer method, 93

schedule amendments, in ICRW, 45–47
Schweder, Tore, 49, 56, 62, 64, 215*n*13
science and politics, connectedness argument: independence myth, 10–11, 19–20; utility factor, 21–30, 33–34, 48, 184; value-neutral problem, 11–21, 31–33
Scientific Committee: non-consensus nature, 182–83; during the *1940s*, 53–54; during the *1980s*, 156–57, 163; during the *1990s*, 234*n*5
Scientific Committee, during the *1950s*: and commission attitudes, 52, 59–60, 64–65; CPUE indicators debate, 55–56; depletion warnings, 57, 59–60, 61, 66–67, 71, 73, 75; Drion's dissent, 73, 216*n*19; quota recommendations, 54–56, 57; Sljiper's dissent, 55, 56, 57–58, 62, 75; structure/working conditions, 48–52, 212*n*39
Scientific Committee, during the *1960s*: and AHSC abolishment, 90–91; commission reception, 111–12; and Committee of Three work, 91–92; depletion warnings, 88–89, 95–97; hunting ban recommendation, 127–28; population estimates, 130; pygmy blue whales, 121–22; quota recommendations, 112, 129–30; Sljiper's dissent, 89
Scientific Committee, during the *1970s*: criticisms of, 226–27*n*32; minke whale debate, 142–43, 226*n*29; moratorium proposal, 138–39, 147–48, 152, 227*n*37, 228*nn*46–47; MSY calculations, 139–40, 142–43, 224*n*9; NMP-based management, 145–46, 149–50; population estimates, 130–31; research proposal, 138–39, 147
scientific finding requirement, in ICRW, 45–47, 79. *See also* Ad Hoc Scientific Committee; Scientific Committee *entries*; uncertainty argument in whaling discussions, overviews
scientific method: and precautionary principle, 7–10; truth problem, 11–21; Type I and II errors, 166–67
Seattle meeting, 84
sei whales: catch statistics summarized,

192–97; in Mysticeti group, 204*n*3; during the *1970s*, 137, 139–40, 142, 147; and technological innovation, 38–39
sei whales, during the *1960s*: catch statistics, 114, 115, 116, 124, 128–29, 223*n*56; FAO recommendations, 126–29, 222*n*40; Scientific Committee recommendations, 103, 112, 114–16, 128
Seychelles, 151
Sixteenth Report, 112
Slijper, E. J., 55, 56, 57–58, 62, 73, 75, 82, 89, 92
Small, George L., 40–41, 104, 215*n*13, 217*n*49
Smith, Theodore Clarke, 207*n*35
smoking example, 33–34
social sciences, utility factor, 26–27, 28
Solyanik, A. N., 117
Sorensen, Heidi, 172
South Africa, 59
Southern Harvester, 104–5
southern right whales, in Mysticeti group, 204*n*3
South Korea, moratorium proposal, 153, 228*n*46
Soviet Union: economic factors, 69, 98–99; moratorium proposal, 152–53; postwar industry development, 43; pre-WWII agreements, 208*n*4
Soviet Union, catch statistics: falsification data, 195–97, 220*n*7, 224–25*n*13; during the *1960s*, 88, 97, 98, 114; during the *1970s*, 138, 143, 222*n*45
Soviet Union, *1950s* quota debates: national-based, 63–64, 70–71, 72; votes/objections, 60–61, 62, 75
Soviet Union, *1960s* quota debates: argument themes, 98, 103, 112, 116, 117, 120, 124, 131–32; national-based, 86, 87–88, 221*n*31, 222*n*39; pygmy blue whales, 120–23; science-based pledge refusal, 111, 220*n*4; suspension proposal, 78; votes/objections, 78, 119, 121, 220*n*4
Soviet Union, *1970s* quota debates: agreements, 131; argument themes, 142; minke whales, 141–43; votes/objections, 141–42, 144, 224*n*9

Spain, 225*n*15
species-based quotas: adoption of, 137; Scientific Committee recommendations, 112, 115–16; in *1960s* debates, 99
sperm whales, 6, 38, 137, 147, 192–97, 204*nn*3–4
Spong, Paul, 170–71, 232*n*96
Stamps, Timothy, 28–29
Stockholm conference, 136
Strong, Maurice, 136
The Structure of Scientific Revolutions (Kuhn), 11, 16–17
superwhale, 169–76
sustainable yield. *See* maximum sustainable yield calculations
Switzerland, moratorium proposal, 152, 228*n*46
symbolic value, whales, 169–75

Takahashi, Junichi, 203*n*1
Technical Committee: moratorium proposal, 227*n*37; quota recommendations, 55, 98–99, 117; structure and responsibilities, 48, 49, 212*n*39
technological innovation, impact, 38–39, 208–9*n*9
10% rule, 147, 227–28*n*41
The Hague meeting, 61–62
theory dependence, 13, 14–16
Tillman, Mike, 151
Tokyo meetings, 56–57, 74
Tønnessen, Johan N., 47, 53, 64, 71, 74, 105–6, 209–10*n*16, 215*n*13, 216*n*27
toothed whales, families of, 204*n*3
truth problem: in science, 11–21, 31–33; and utility factor, 21–27, 33–34
Tverianovitch, V. A., 103, 117
Type I and II errors, 166–67

uncertainty argument in whaling discussions, overviews: during the *1950s*, 36–37, 47–48, 51–52, 68–69; during the *1960s*, 68–69, 79–80, 100, 102–4, 108–10; during the *1970s*, 140–41, 153–54; summarized, 3–7, 177–88. *See also specific topics, e.g.*, blue-pygmy whales; quotas *entries*; science and politics

United Kingdom: catch statistics, 97, 98; economic factors, 43–44, 67, 98; pre-IWC meetings/agreements, 41–42, 208*n*4, 210*n*18; in protectionist movement, 174; public opinion poll, 233*n*101; resistance to postwar newcomers, 41, 42–43; sales of ships, 78, 97; World War II whaling, 41, 209–10*n*16

United Kingdom, *1950s* quota debates: argument themes, 60; IWC withdrawal notice, 71–72; national-based, 63–64, 70, 72, 74; votes/objections, 58, 59, 61, 62, 75

United Kingdom, *1960s* quota debates: argument themes, 99; committee proposal, 79; national-based, 86–88, 221*n*31, 222*n*39; pygmy blue whale, 101, 104–5, 120–21; suspension proposal, 77–79; votes, 99

United Nations, 122, 136, 137, 139, 171. *See also* FAO (Food and Agriculture Organization)

United States: aboriginal whaling distinction, 203*n*1; and Committee of Six, 124–25; environmental protection legislation, 223–24*n*6, 225–26*n*18; moratoria proposals, 136–37, 139, 140, 141, 151–52, 153, 227*n*37; NMP proposal, 146; pre-IWC meetings/agreements, 41–42, 210*n*18; public opinion poll, 233*n*101; in *1950s* quota debates, 59, 61; in *1960s* quota debates, 99, 119, 125

utility factor, generally: in cancer cure example, 34–35, 177; and precautionary principle, 29–33; in resource exhaustion, 212–13*n*49; as scientific criteria, 21–30, 33–34, 48

value-neutral problem: in science, 11–21, 31–33; and utility factor, 21–27, 33–34

vetoes. *See* objections

votes: during *1950s* quota debates, 56, 58–59, 60–62, 75, 90, 214*n*79; during *1960s* quota debates, 78, 99, 111, 119–20, 121, 128, 220*n*4; during *1970s* quota debates, 141–42, 144, 224*n*9

Washington Conference, 42–43, 210*n*18
whale oil, production statistics, 208–9*n*9
whaling olympic, impetus for, 40–41
White, Ben, 171
Willem Barendsz, 105–6
withdrawals from IWC: during the *1950s,* 62–63, 71–72, 74–75, 76; during the *1960s,* 77–78, 87–88, 216*n*27
World War II, whaling industry, 41, 209–10*n*16
World Wildlife Fund, 144, 169

Young, Oran R., 211*n*34